Stuart Baker is a rare breed—someone wh practices the very principles he teaches. stories that, if taken to heart by both bu ...., van prevent many heartaches and frustrations, and even lawsuits. Having seen these principles in action—as well as having experienced other situations where either my contractor or I ignored these principles—I can attest to how the small investment in buying and reading this book can pay back huge dividends in satisfaction with work done well, as well as the potential for positive and enjoyable relationships between a contractor and an owner. Highly recommended!

Daniel Stone

Organizational change consultant

Stuart's book is a gem that I can highly recommend. It is full of wonderful stories and very practical information about how to make the construction process a wonderful experience for all parties. But beyond that, it is a great book for anyone looking to improve their skills of dealing with other human beings during difficult processes. Go out and buy it right now!

Brent Darnell

Best-selling author and expert on emotional intelligence

I thoroughly enjoyed Stuart's book and his ability to relate such wisdom to relationships in the construction workplace and life in general. It is a one-of-a-kind work gracefully diffuses, with mutual respect and understanding, the potentially volatile and misunderstood relations between the building contractor and customer. What stands out the most for me, he captured the heart without fanfare or embellishment and mirrored through personal experience and expertise- a much-needed "conscious cooperation" for all relationships.

Jeri Costa

Personal coach and energy healer

CONSCIOUS COOPERATION

HOW TO CREATE SUCCESSFUL CONSTRUCTION PROJECTS

STUART BAKER

iUniverse, Inc.
Bloomington

Conscious Cooperation
How to Create Successful Construction Projects

iUniverse books may be ordered through booksellers or by contacting:

iUniverse
1663 Liberty Drive
Bloomington, IN 47403
www.iuniverse.com
1-800-Authors (1-800-288-4677)

Because of the dynamic nature of the Internet, any web addresses or links contained in this book may have changed since publication and may no longer be valid. The views expressed in this work are solely those of the author and do not necessarily reflect the views of the publisher, and the publisher hereby disclaims any responsibility for them.

Any people depicted in stock imagery provided by Thinkstock are models, and such images are being used for illustrative purposes only.
Certain stock imagery © Thinkstock.

ISBN: 978-1-4759-4461-7 (sc)
ISBN: 978-1-4759-4462-4 (ebk)

Library of Congress Control Number: 2012914931

Printed in the United States of America

iUniverse rev. date: 09/21/2012

CONTENTS

INTRODUCTION

If conversation at a cocktail party or a coffee shop turns to construction projects, there is often cringing and at least one horror story. Whether the subject is a nightmare of a remodeling job or a public sector project mired in lawsuits and sitting uncompleted, all too often construction is contentious. I hazard to say that most people either have had a bad experience with a construction project or know at least one person who has.

What are the factors that contribute to this contentiousness? Why do so many construction projects end up in some form of serious dispute?

My career in construction is now in its fourth decade. I was always fascinated with plumbers, carpenters and electricians around the house when I was a boy. My friends and I explored new house sites with eagerness. Little did I know that I would end up building custom homes and carrying out myriad remodeling projects.

In addition, little did I know that my dislike of conflict would lead me to study my working relationships in construction and the roots of conflict, eventually become a mediator and ultimately write this book.

As my involvement in the field of construction grew, it became apparent that the working relationships were hugely important. I did some simple observation, questioning and deduction to see that customers love to feel well cared for and listened to. Unclear detail and description of a job can easily lead to conflict. Unasked questions can lead to assumptions and conflict. Poorly written or nonexistent contracts can lead to conflict.

These factors are in a sense the business end of clarifying a construction project, but there is much more to successful projects than clearly written agreements and even excellent work.

The working *relationships* are critical, and often paid little attention. I found that a *conscious intention* to please my customers and have cooperative, cheerful jobsites is a huge plus. I found that to keep them informed, and also check in with them repeatedly, go a long way toward forming and maintaining pleasant, mutually satisfying relationships.

It became clear that bringing into construction a deeper human side that includes clear and direct conversations, humor and fun, and a little sharing of personal life brought in Technicolor to the facts and trades related to building and remodeling.

There is a story in this book of a woman who became a customer of mine. First we did a very small job for her, and then we built her an entirely new house off the original main floor frame. Her friends warned her that she would hate me and probably end up in legal battle with me, because supposedly that is just the way things are.

Not only did this scenario not play out, but she enjoyed her experience so much that she decided to try to help others carry out projects similar to hers. Her house incorporated an integral suite for her handicapped mother, and she got thoroughly engrossed in making numerous changes that she saw as improvements for either her mother or her whole family. The house was later featured in an article in Fine Homebuilding magazine, and my customer had her own sidebar in the article.

This book is about concrete, tried and true steps contractors and homeowners can take to build and maintain successful working relationships from the beginning. It is also about how to deal with conflict that may arise, even with the best of intentions and efforts.

This book is also about working together in harmony and even enjoyment. What a novel idea for construction!

I hope this book adds something to your life. All the people who participated in the story that led to writing it have added something to mine.

Stuart Baker
Virginia, 2012

CHAPTER ONE

The Roots of "Conscious Cooperation"

A seventy-two year old man called me several years ago to ask if I could review a partially completed renovation project at his house. He wanted me to estimate the value of the work completed to date, estimate the value of the remainder left to finish, and assess the quality of the work done so far.

During the course of my time with him he said to me about the builder, "I may be old, but I really want to hurt this guy. I have hurt people before, and I am ready to do it to him." I did not doubt him at all. Yet I told him I was hoping he could contain himself, and maybe telling me the whole story would help relieve some of the pressure inside him.

How did a seventy-two-year old man get to such a drastic position?

The story was awful. According to this man, the builder kept saying he needed more money over and above the estimated contract. He left previously remodeled sections of the house unprotected from violent storms, causing ruin, and the whole thing had been dragging on for about two years and was still uncompleted. The entire project should have taken several months. When I met the man, he had already paid the builder more than twice the amount of the original contract, aside from valid extras that were mutually accepted. He also said that the builder demanded money for windows he never actually ordered for the customer.

I asked him how he could have let things get to this state. He said that his wife had been very ill, as had her father, who may have passed away; I do not remember all the details now. The point is, he said that the emotional and health crises were pretty overwhelming, and he kept fearing that if he did not continue to give the builder the money he asked for, the builder would quit, leaving him in even more of a mess.

He said the builder had continual stories and excuses and tried to turn it all around to seeming as though all the conflict was the customer's fault. The house was located by the ocean, and the builder kept telling his customer that his house was worth a pile of money because of the location, so what did he have to complain about!

I really felt for the guy. Not only had he experienced emotional and financial nightmares with his project, in addition I had to honestly tell him that some of the work done was not very good.

That was one horror story from the property owner side of the construction relationship. Here is a story from a builder:

Several years ago the main lumberyard that I use for my building supplies agreed to put on an evening for me with their most valued commercial customers. It ended up being a pretty high-powered group. I knew most of the builders who came to the evening. Some I invited personally. The one who gave me the most resistance did come, but he told me he did not want to come if I was just trying to sell something.

I assured him that the evening was for the guys who came; they would be the stars of the show. What I was looking for was how the contractors who attended the evening were doing with their relationships in their construction businesses; what was working, what was not, and what were they needing or looking for. What was their main source of pain, and what might help?

After introductions and a quick delivery on what the evening was about, the builder who finally really broke the ice in open discussion was the one who gave me the resistance about attending. He spoke for about fifteen minutes about a customer from Heck. This builder is honorable. I have known him for years. He is highly motivated and conscientious. He is good with people. He has a friendly personality. He does careful work.

He said that his customer (I believe an attorney from New York) acted displeased no matter what he did and would not pay money legitimately owed. Whatever their agreements were about finishing final details, the customer always said he was displeased and came up with new demands. The builder was at his wit's end. He clearly experienced some relief simply sharing this awfully painful story with us. Then, at the end of the evening he said, "Hey, this meeting was great! I hope we can continue having them!" I felt for him, as did I think everyone in the room.

There are horror stories on both "sides"—construction people and property owners. You will see that what I contend, and offer training in, is not only that there do not have to be firmly defined "sides," but that holding to such a stance can greatly harm the chances of success for any given project. In general, builders frequently are considered to be on an ethical level equivalent to the oft-maligned car sales people. Right or wrong, this rating is not very high, to say the least. There are strong impressions and attitudes that construction people just cannot be trusted, and that if you stop standing over them for one minute they will take advantage when there is the slightest opening.

Conversely, many builders have stories of misery they have experienced working with certain customers that seem to make them look like the poor victims of scheming, dishonest, rotten property owners. I am sure that sometimes this is the case, but have found over time that the truth generally lies in a place where there is not an evil demon on either side forever tormenting the utterly innocent one on the other side.

The truth is usually somewhere in the middle. The way we see things may be heavily colored by our upbringing and our world experience. We get messages imprinted into our minds that often put an unintentional spin on daily events and predispose us to react in a particular way, which may be pretty far off the mark. Aren't you amazed sometimes at how someone gets from point A to point B in their mind, when you can see nothing at all that led to that journey?

A simple case in point:

When I was finishing college in the Boston, MA area, I drove a cab for the Cambridge Yellow Cab Company. One day I picked up an older woman for a trip of three blocks, if I remember right. Before she even had the door closed she started ranting at me about dishonest cab drivers, how they always try to take you out of the way to jack up the fare, how they will use any trick they can to soak extra money out of you.

By the time I had gone the three blocks to her apartment building, she was foaming at the mouth, yelling at me. Spit was actually coming out of her mouth. I had not said one word. She finished her diatribe with, "And just for that, you're not getting a tip, either!" I, too, was one of those miserable, dishonest cab drivers, even though I drove her down one street for three blocks and did not open my mouth. Obviously, I never forgot her!

....

There is a well-known expert on buying, renovating and flipping houses. She is a woman who was highly successful in a depressed, really challenging locale during poor economic times in general.

Her level of success was especially noteworthy for where she did her flipping during these times.

I attended a seminar where she was one of the key presenters. I liked much of what she said. She advised the participants to always offer more value than similar houses for sale in the same area. She said you do not have to spend a lot of money to do this, but to make your house stand out. She would also arrange for financing and pre-qualify her buyers for mortgages. She helped applicants fill out the paperwork. She covered all the bases. She was thorough and confident. I was generally impressed.

Then she said if we took further training with her, she would show us "how to keep contractors on a very short leash." That is where she lost me!

An attitude of keeping contractors and their subcontractors and employees under control, as if it is an "us against them" battle, is, to me, exactly what is often wrong!

On the other side, I have known talented, conscientious building contractors who had the attitude that ALL customers would try to get away with something and take advantage if they could. One good friend fit into this category. I told him that I was getting along great with most of my customers, and I had valuable relationships with them. I also told him that I had come to expect to have this kind of relationship, and I was willing to do my part to make it happen. He said that he simply did not have those sorts of relationships—as though it were an inevitable result of who his customers were.

His situation was really a pity, because he was highly talented and honest. He had just convinced himself that he would always have trouble with customers—and sure enough, he always did.

We get what we expect and put out.

John Hochbaum from Contracting Trust told me that it is typical that every year complaints about builders and their work are in the top two categories of consumer complaints in the United States. That is awful!

How has that developed into such a common state of affairs? Many people seem to accept that having construction projects performed is always going to be worse than having root canals done with no Novocain. That being the case, it's a wonder anyone ever has any construction done at all!

....

Years ago, I had a wonderful customer who hired me to do some simple remodeling to a half bath at her second home. At the end of the project, she was very happy. We sat and chatted. Eventually she mentioned that she had blueprints to turn her seaside California-style bungalow on Cape Cod into a completely new two-story house. Did I want to see the blueprints? I looked at her in amazement and said "Sure!"

There were significant mistakes with the plans. She asked if we could proceed ahead without an architect. Again, I said "Sure!" I told her I could get the problems worked out and the plans redrawn. She started to get highly excited. Then she told me that all her friends warned her that she would end up hating her builder, no matter what.

I told her I did not expect that to happen, and my intention was to make her very happy. She said she had a good feeling about me, and she was eager to move ahead. What an unexpected development from doing some relatively minor work in a powder room! We did move ahead with this ambitious project.

A big part of what was most interesting about the project was that she was incorporating into the house a handicapped-style suite for her elderly mother. My customer became so involved in the project that during the course of it she constantly came up with ideas and questions about making her mother's suite better and better in terms of its function. She left me many messages at two or three in the morning, because she became so wrapped up in trying to take care of her mother the best she could.

We logged forty-five work change orders by the time the project was completed. The house eventually appeared in Fine Homebuilding magazine, and my customer had her own sidebar. She loved her involvement so much that she decided to become a consultant for people dealing with aging parents and altering their homes to incorporate their parents into them.

All this from a woman who was told that she would end up hating me.

....

I have been a building contractor and carpenter for over thirty years. It was not too long into my career that I started to get a hint of just how important the human relations end of the mix is. The relationships between customers and contractors, contractors and subcontractors, contractors and architects, contractors and realtors, contractors and engineers, and all of these relationships, eventually stood out as being every bit as important as the physical work at hand.

Every bit as important, yet so often so poorly attended to!

I started to review how my jobs went. What were the more successful projects which, not surprisingly, also had successful and often mutually rewarding relationships? Where was conflict? How did conflict arise? *What could I do to reduce and even eliminate conflict*? What did customers have a right to expect of me? What did I have a right to expect from them?

As I mulled over these questions and others, I had the inner nudge to, from that day forward, make blunt discussion with new and old customers a matter of primary importance right at the beginning. Before ground had been disturbed or one nail driven.

Before there was even a contract. In time I realized it was important to ask about things no one ever tells an aspiring contractor to explore: their fears, their wishes, what their house would mean for them, how they thought their new home or remodeled home would change their lives, and so on. I needed to find out their expectations, hopes and priorities.

At first, I was somewhat nervous venturing into this territory. How would people react? Would they say, "How dare you ask us questions like these? What kind of nonsense are you talking? Let's just focus on getting this job done and what our contract is." They might feel affronted. They might tell me to leave.

What in fact happened is that when I asked plenty of direct questions before there was even any contract, most people looked at me slack-jawed and then said something like, "This discussion makes us feel very safe. We never dreamt that our builder would bring up things like these questions and issues."

In addition, many of them became friends along the way and remained so afterwards. We had dinners together. Some asked me to stay for dinner after a day of work and then told me to call my former wife and invite her, too. They would ask during the project if I needed money. Could they help me in any way? One friend who worked with me on and off said he had never seen customers chase the builder to hand him money. He said, "How do you do it?!"

By then I knew I was committed to giving my best to my customers, and I let them know this, along with asking all the direct questions. My work started to come almost exclusively from referrals and repeat customers. I had to make a concerted choice to make this commitment. I knew I had opened a door to greater responsibility and transparent integrity. I would have to consistently deliver on my promises to succeed in taking such extraordinarily good care of them!

A number of customers told me they knew I would always have their best interests at heart. They trusted me to give them fine work, to advise them of unexpected situations and make suggestions for extras and changes they might like. Achieving this level of trust became a goal of mine on every project. Frequently it was enjoyable, although it was exacting, too.

Sometimes I had to deliver news that they might not embrace with great joy, like discovering the need to reconstruct part of a house that was assembled poorly in the first place, or some other unexpected development. Yet I spoke honestly, and typically my customers took in the news, asked questions and then asked how best to proceed ahead. Maybe it was a sign of the success of my more open approach to customer communication, but I never had the feeling that they wanted to shoot the messenger.

Again, I found that forthrightness was the way to go. Tell the truth, give options, and leave my customers the space to have their reactions and responses. I became comfortable with silence. Silence at the right time can be more powerful than words.

My customers often felt like teammates. Some took assignments from me, some offered to help if they would not be in the way. I once had one of my frequent customers spray-prime some cedar boards. She had never done such a thing, and she did pretty darned well.

Another customer, newly retired, agreed to do the drawings to get a renovation permit, with my coaching. He did very well, too.

I loved it, and you can bet that I wanted to take excellent care of my customers in return. We often had a mutual admiration society, and could not do enough for each other. Many customers told me I worked too hard. One couple was surprised to see me working on the weekend, and asked why I was doing that. I told them that a crew was coming on Monday to do an important piece of work, and some prep needed to be done ahead of time. This particular couple became friends of mine, and over the years had me do perhaps five projects on their house. Nevertheless, that day, when I was still "just" their contractor, they shook their heads and said I should take some time for myself. In time, they also repeatedly invited me to sail with them. The relationship grew into what seemed to be a friendship that occasionally had work thrown in. This was an interesting development that I had to get used to, and allow myself to enjoy and appreciate.

Imagine, my customers telling me to go home and take care of myself! They were genuinely concerned about my well-being. What kind of a pleasant dream is that for a builder? As we came to know one another better and they invited me on their boat, they often told me to call my wife and invite her for dinner on the spot. I had to do a dance to balance the emerging friendship with the working relationship. It was a nice dance, though.

CHAPTER TWO

Key Questions and Statements

So, what did I learn to ask? Here is a sampling of the type of questions that became important for me to ask new or potential customers:

1. **What is most important to you about your project? What are your priorities?**
2. **Do you have any fears or hesitations around doing the project?**
3. **Is the money a scary factor to you, or are you comfortable with it?**
4. **Do you have prior experience with having building or remodeling projects done? If yes, what was your experience?**
5. **Do you have any related horror stories of your own, or from people you know, that might be influencing you?**
6. **What do you expect of me?**
7. **Do you trust me?**
8. **Do you think you have any prejudices or tightly held impressions about construction people that could influence our relationship?**

9. **What is your best picture of how your project would go?**

10. **What are your hopes around the project?**

11. **What does your house mean to you? What is it for you? Do you entertain frequently, do you love your privacy?**

12. **Are you satisfied with the blueprints or plans you have—with all due respect to the architect or designer?**

13. **Do you have any hesitations about doing the project?**

14. **If you need financing, do you have it in order?**

15. **Have you picked out your fixtures and agreed on details?**

16. **What should I be able to expect from you?**

17. **As a couple, do you make decisions together?**

18. **With whom will I be talking most of the time?**

19. **Can you and I agree to have a partnership, to be a team?** Because that is how I want to work. I am on your side, and I want you to be on my side, okay? I am here to give you the best of me, and I want you present with me.

20. **Is there anything you would like to know about me that you do not know?**

21. **Is there anything you would like to ask me?**

22. **Are you comfortable enough to proceed ahead?**

23. **Is there anything at all here that is being left unsaid or unasked? I really want to know.**

24. **Construction projects often involve some degree of tension.** There may be disagreements. There may be misunderstandings. Memories may fail, even if we are as detailed as we can be. In other words, there may well be some conflict. You are spending what may be a very significant amount of money, and you are putting yourself

in my hands to deliver a quality product that you can enjoy for years. I understand how you might have trepidation about letting me into your lives like this and giving me this amount of money. *With all these factors in mind, expect some tension.*

Years ago I built a custom home for a marvelous couple. They were both really likable, very bright, and excited about their home. I gave them the above introduction before we started and said that they might well reach a point some months into the project where they were sick of it all and wondering why they ever hired this stranger who they were forced to have this intimate daily relationship with while he was building their house.

When we were all done, and we got along great, the lady told me that four or five months into the project they did hit the wall I said might show up. They were exhausted and strung out, even though overall we were doing very well together. Then she remembered that I had warned them about just such an experience. She reminded her husband, and she told me their tension just melted away. I was glad I had delivered that one-minute part of my speech!

25. **I want you to know that if I tell you something I mean it, and I want to be able to expect that the same is true for you.** I want directness and honesty between us. If we are going to work together, we need to work with the truth, and we need to trust one another. If we do not trust one another, I do not want a contract together, and I will not do the job. How does that sound to you? Can we agree on everything?

....

Intention is very important, too. My intended purpose with these questions is to do my best to establish the solid beginnings of an open, cooperative, trusting relationship. I find that most new customers experience visible relief with this kind of conversation. Sometimes they sigh and relax into their chairs. You can sometimes see them decide to open the door to themselves wider. They invite you in. They show more of themselves.

This is the basis of successful relationship, which makes a project better all around. If your customers trust you deeply, and are grateful they have you to carry out their work, your project and your life are going to be much more enjoyable. Your customers will tend to be more understanding of what it really takes for you to deliver their dream. You might well find some of your customers asking you how you possibly juggle all that you do day after day.

WHAT IS BEHIND THESE QUESTIONS AND STATEMENTS

When I pondered the roots of conflict, and in addition took training in mediation skills, some key observations stood out as identifying what frequently fertilizes the ground for conflict. Here are some of these observations:

1. *Assumptions.* If I decide I "have your number" or understand you, without checking out the validity of my assumptions, chances are I will be quite off base in some way. Statements like "But I thought you meant . . . ," "Oh, well this can only mean . . . ," and "But I assumed . . ." are battles in the making. Major lawsuits have originated based on erroneous assumptions. *Asking direct and clarifying questions is one of the best cures for the epidemic of "assumptionitis."*

2. *Clarity.* Attend to the details, and get them out on the table! Be detailed in discussion, and in your specifications. Root out the assumptions as best you can. Ask clarifying questions. Do not be afraid to look like an idiot because you are asking plenty of questions. I would much rather appear to be an idiot who is armed with correct information and lots of answers because I was not afraid to ask. *Great clarity also eliminates a lot of interpretation.*

3. *Put it in writing.* There are those who say that they do not need contracts. They do not need write-ups. They say that if they cannot trust a handshake agreement, they do not want to do business. In an ideal world, this would be a wonderful way to approach even the most complex project. Yet I have found that even where there is great trust, respect and even affection, human memory is prone to lapses. I know this well! A good written proposal and a good written contract, which can be easily based on the proposal, serve as a record for all involved of what is to be done, how much it is estimated to cost, the timetable if there is one, agreement on payments, and so on. This is a simplistic definition, but my experience taught me the hard way: *make detailed write-ups of everything agreed to, signed by all the key people involved.* Review the details together. Ask if there are questions. Ask if anything was overlooked. *Asking questions clarifies like nothing else, and lays the foundation for a smooth and harmonious relationship.*

4. *Focus written proposals and contracts on the work, and not on legal protection and recourse.* If you set an initial tone of legal guardedness and recourse, you begin with two strikes against a potentially effective and enjoyable working association. If instead you focus on the work

ahead from a desire for clarity, you open the door to a practical cooperation based on trust.

5. *Allow for and anticipate conflict.* Tension is built into construction. Nerves get frayed. Even the best of intentions get thwarted. Delays happen out of anyone's control. Weather happens. People may let you down. Things break. Unexpected things happen. *Human imperfection keeps showing up!* Misunderstandings happen even in great relationships. Acknowledge right up front that this could well occur!

6. *Agree to try to resolve conflict on your own, but if you cannot seem to, turn to mediation for the next line of defense. Do not turn immediately to attorneys in an adversarial fashion.* People often leap so quickly into legal action, they do not even think about it. Many people think this is the only way to go. That is not true!

Adversarial legal action is miserable and costly, and in the long run, everyone might end up losing in one way or another. More on this subject later.

In mediation, both parties involved have the opportunity to resolve conflict on their terms with the help of a third party who is not emotionally tied into the situation.

Here is an unusual mediation story from my own experience:

My good friend Jeff Oppenheim, who is a compassionate attorney and a wonderful mediator, and the one who convinced me to take initial mediation training, referred an interesting dispute to me several years ago. A nice couple had been in unresolved conflict for months with a granite and tile contractor who had contracted to install new countertops and new flooring in their kitchen. He had also agreed to get some related carpentry work done.

A series of disputes and complaints had mushroomed into an unfinished job with festering anger and no communication. The couple was angry. The contractor was angry. The work was not finished. The money was not settled.

I met with the couple and saw their kitchen. They told me their side of the story. There was a real stalemate. The wife was so upset with the contractor she could not even try to talk with him. We discussed what they wanted to propose to the contractor through me. They did not want him back in their house. They were very decent people stuck in the middle of an all-to-common scenario in construction projects.

The contractor did meet with me. Our talk went pretty well until it came time for me to make the financial proposal from the homeowners. The contractor's face immediately shut down when I told him what the homeowners thought was fair. End of discussion.

I reported back to the couple, and that was that for now. They repeated that they just did not want the contractor back in their house. They said they did not trust him to finish the job in good order and to their satisfaction. I was disappointed that we did not get any further, but I could not do anything more to bring the two sides together.

Perhaps three weeks later the husband called me on a Saturday. He said that he and his wife had rethought everything, and they would be willing to allow the contractor back into their house for completion of the project under supervision. If the contractor would agree, would I take on the role of supervisor and final judge of the work? I was encouraged by the call and said I would be happy to contact the contractor and see if we could make progress.

The way this couple had opened themselves to new ways of looking at their painful situation impressed me a lot. The husband said they wanted to get the whole thing done and move on from what had become a most disturbing phase of their life. I could feel the husband setting part of his ego aside and stepping into a broader place of intention born of goodwill. *This is actually a key component of the healing process of mediation, and he had ventured into it on his own without help.*

I reached the contractor, and he agreed to sit down with the husband and me. The small yet formidable wife was so upset with the contractor that she knew she could not be anywhere near him, and she did not want to sour the new possibility of resolution. She too, wanted resolution. She has a great spirit, but in this case, that spirit and feistiness were focused against the contractor. For his part, the contractor also did not want her around.

We three men met at the couple's home. In a little over an hour we reached an agreement on the work to be completed and the final money to be paid out. Both the husband and the contractor also quickly assigned me the role of supervisor and ultimate judge of the quality of the work. I pointed out that both of them would have to be happy with my judgments, and that I am a pretty fussy builder. They consented. No question. We all shook hands and set a date to begin what looked like two days of work to finish the job.

With the consulting help of my friend Peter we also resolved how to deal with an important structural issue: how to support a prominent overhanging granite countertop. I ordered some decorative posts that I modified to fit the situation. This challenge had been a raw sticking point, and the couple was happy with the solution.

Day one of the work went along very well. The wife was careful to stay invisible. I could see she was trying hard to do her part to support the new positive process. I checked in and out during the day.

Day two went fine during the morning. I had suspected that the carpentry subcontractors who had traveled from Boston to Cape Cod might not have all the tools they would need for the day. My house happened to be about two miles up the road, and I brought tools I thought the carpenters may not have. It turned out that we did need those tools.

I started to make a piece of wood trim, and the lead carpenter from Boston seemed to quickly decide that I had a good idea what I was doing, and he jumped right in to help me. This was a good sign!

I was spending quite a bit of time at the job now on that second day. At one point I left to get some supplies. I was gone for maybe twenty minutes. The head contractor had not been there when I left—he had not been there at all during the day—but he called me while I was out. He was very upset. He was at the job and said the wife was demanding to have things done that were not part of the agreement. He said he was fed up and ready to leave. I said, "Get away from her. I will be back in a few minutes."

We both hung up, and a minute later, the wife called me. *She* was highly upset and claiming that there were five construction guys in her driveway, all angry at her. What I have not said yet is that the contractor and all of his men were from another country. They were angrily speaking about her in their native tongue, which further widened the gap and contributed to the wife feeling intimidated. I told *her,* "Get away from them. Get in the house. I will be right back."

In two minutes I was back at the house. Sure enough, there were the five men huddled in the driveway. I plunked myself in the middle of them. The head contractor said with great upset that he had had enough and was ready to leave.

I said to him, "Two or three more hours of work and you are done. You will get the money owed you. Don't blow it now! I will work with you. We will get it done together." What I had just realized was that I had to stay there and work right along with them to keep the peace.

I also told him that if I felt the wife was being unreasonable I would tell her. If I felt she was justified I would tell him and his workers. This soothed them. I strapped on my work pouch, and got ready to dive in.

Having a quick meeting with the lady of the house, I did see that part of what she wanted done was in fact not part of the agreement. She readily said okay when I told her. I told her that we wanted her to be happy with the work, and when we were done she would make her final inspection. I told her we *planned to make her happy.* She was satisfied. Again, I could see her doing her best to play her role in getting it all done. I pointed out to her that two men were cleaning the tile floor with *toothbrushes!* I said you cannot get much more detailed than that!

In another hour or two, we had the job completed. The wife made the final inspection and was *happy.* She asked what I thought of the work. I said that the men had made their utmost efforts to finish everything in good order, and I was satisfied. This was all she had to hear.

She called her husband at work to report to him and discuss the final payments. He got on the phone with me for a minute and agreed to the final figures, which included an extra or two. He wanted his wife to be happy, and if she was happy, he was ready to move on. There is a saying that there are two words for a successful marriage: "*Yes, dear!*"

All ended with handshakes and smiles, and final money paid. I drove away shaking my head over the highly unusual and comical dance of the day, and very grateful for the resolution. Moreover, I was grateful that I had had the opportunity to play a little role in resolving this painful scenario. The fact is, I love being able to be part of helping to bring harmony where there is strife.

I want to note that the multiple roles that I took in that particular situation are not standard mediation practice at all. Yet, the customers did hire me to play those multiple roles, and I felt comfortable doing it. Both sides left me the full responsibility of coordinating the work and being final judge of whether the work was satisfactory. In the beginning, I never had any intention of strapping on my tool belt. That was an in-the-moment decision that worked. I was sure that things were going to blow wide apart if I did not take that step.

What I did was definitely "Wild West" mediation, certainly not by the book. Many mediators might cringe over my actions, but I knew I could help finish the job in good order while deepening my bond with the contractor and his men doing the work. The woman of the house was also happy to see me participating in the labor. It clearly did become a team effort, and everyone seemed to just kind of merge in that direction.

CHAPTER THREE

Where Does It Hurt?

The marvelous business and marketing coach, Mark Silver, says in his highly insightful book *The Heart of Business*, that successful marketing is about identifying the pain of your real and potential customers and then stating how you can give them relief. Then you offer that "cool, refreshing drink," as Mark says. I think this is profound, and so simple! It is getting down to the truth, to what really matters.

You can gather all kinds of demographical information and put it together beautifully, but if your customers don't feel that you can and will take good care of them, if they don't feel that you understand what is important to them and are making that very thing important to you too, to me you are always basically playing catch-up. *It is crucial to establish rapport and mutual trust at the beginning.*

So what is the pain that often arises for contractors in construction relationships? What is the pain for property owners? Is this pain ever the same, or related?

If we can uncover the truth in answering these questions, we are well on the way to knowing how to set the stage from the outset for successful, *mutually beneficial* working relationships in construction projects.

What are the needs and wishes of property owners, and where can they experience pain?

Whether your project is remodeling a bathroom or building a $5 million home, here are some common wishes and needs listed below. Commercial construction is a bit of a different animal than residential building and remodeling, with some different priorities, and we will touch on that, too. Meanwhile, here is the list of what I see as important common wishes and needs of property owners. This is based on my own years of experience and some interviews:

- Quality work. Most people care about having quality work done.
- Fair price. If your priority is the absolute lowest price, you can expect that you will get what you pay for. *Fair price* does not necessarily mean the lowest price. Lowest price and quality do not often go together.
- Courtesy and respect. People like to be treated with courtesy and respect. See the Golden Rule.
- Understanding. Customers like to feel understood. This is crucial for carrying out a construction project. If the contractor does not understand the homeowner and the plans, there is big trouble. People also like to feel *heard*. You know the difference when someone really takes the time and space to listen to you well, with genuine focus, and when someone interrupts you and/or just cannot wait to jump in with what *they* want to say. The difference is immense, both in terms of how it is experienced and in terms of effectiveness of the communication.
- Timeliness. When jobs go way over the time scheduled, there is usually trouble. If there are unexplained absences and delays, there is usually trouble.

- <u>Honesty</u>. People want to feel they are being told the truth! This includes being true to your word. Moreover, if something is not going well, or there has been an error, own up to it—no matter how uncomfortable it is, and the sooner the better. This disarms suspicion and assumptions faster than anything else.
- <u>Intention of cooperation</u>. People like to feel that there is an underlying tone of harmony and cooperation. Cheerfulness is included here, too. If there is a definite sense of disharmony and distrust, everything is soured, and everything is affected.
- <u>A sense of caring</u>. A customer likes to feel that the contractor and all workers involved *care* about the project and their work. A sense of pride in workmanship comes under this category. You can *feel* a project that has been finished with care and pride in workmanship. These qualities ooze out of the pores of a carefully done job. Even totally untrained people often pick up on this tone.
- <u>Communication</u>. Keeping the owners informed of what is going on, both good and not so good, is often overlooked, yet it is so important! Many lawsuits have arisen out of poor or non-existent communication. I also learned this the hard way. After a while, it became pretty easy for me to grab the phone and call my customers to let them know about either good progress or a turn of events that I knew would make them happy, or about unexpected complications that had to be dealt with.
- <u>Respect</u>. The owners would like to be treated with respect by the contractor and anyone working with them. They would like their property to be treated with respect. Don't they deserve this?
- <u>Genuine promises</u>. How many stories of construction projects gone really sour do you hear that seem to focus on

broken promises? It seems to be <u>so</u> important to property owners to have promises kept. It is important to honor one's word. *Furthermore, be careful what you promise!* Be careful what you are committing to. In addition, for God's sake please do not "tell people what they want to hear," knowing full well you don't intend to carry through!

- <u>Cleanliness</u>. I have known good artisans, likable guys, who seem to take pride in making messes and leaving them for someone, *anyone* else to clean up. Many customers have commented that they really appreciate cleanliness and orderliness on the job, and went out of their way to acknowledge that I often went beyond their expectations in cleaning up. A few times, I have done multi-week, multi-month jobs for one customer, and I make a point to clean the lawn where much of the cutting has taken place. I rake and even vacuum. A little extra effort is usually deeply appreciated and it really registers with people. This is so simple and can be quick to do.

Here is a story about job site respect:

Some years ago, I had an addition/renovation job in an exclusive community on Cape Cod. Some very well known people lived there. My customers were a delightful international couple. They were charming and humble. They always treated me with great respect and appreciation. I did the same.

One thing I am a normally a stickler on is keeping a job site neat and orderly. I try to be conscious of the neighborhood if houses are around, too. There was a portable toilet at this site. One day I saw one of the members of the framing crew, who I had known for years, go *behind* the portable toilet to relieve himself. I think I actually rubbed my eyes—was I really seeing this? He would be visible to at least two neighboring houses behind my customer's.

"What are you *doing*?" I asked, not masking my disbelief.

"Taking a leak," he said.

"What do you think the porta-potty is for?" I asked.

"Oh, those things smell. I never use them," he replied. This guy was lovable, but not the classiest guy in the world.

"The thing was just cleaned!" I let him know. "Just use it, will you?!"

This same guy would sit for break time or lunch time ten feet from a dumpster and throw his trash on the ground. I guess he figured the maid service would be coming through later. He and his crew-mates *all* did this at another job I hired them for. It was windy as heck, and their wrappers, bags and coffee cups ended up everywhere. Some of the guys could have just about reached out and touched the dumpster from where they sat. I told their boss, a good friend, that if it happened again I was back-charging him for cleanup. He just said, "Got it." It did not happen again.

Sometimes the cave man attitude of "I am construction guy. I am rough. I am loud! I make mess!" is pervasive. They can be very proud of being crude, like it just comes with the territory. Oh well. It can be both funny and maddening at the same time. You can love them while also wanting to whack them with a bigger board.

....

The same highly skilled framing boss was toward the end of his own fine custom project some years before when the project designer, who did not get to use his usual builder, decided to try to turn the homeowners against my friend. To add to the complications in this situation, my builder friend had become friends with the customers. In any event, the designer had his way. My builder friend's go-to response was to clam up and say nothing when wrongfully attacked. The customers took this as an admission of guilt. My friend ended up taking out a bank loan, after legal wrangling, to pay the customers back for some of the work he had already done. God, what a mess! Communication along the way, and better communication skills in general, probably would have helped even the playing field and lead to a more equitable outcome. My friend was soured on the entire industry for a while and really had to heal from this painfully nasty experience.

I think that most property owners would probably agree on the above points. Of course, we all might have different priorities, but these points are pretty common and understandable. I am not saying I have absolutely included everything, but I think the list is a fair representation.

....

Here is a story about making promises:

Some years ago, before I met Ted who has been my electrician ever since, my regular electrician suddenly decided to do away with his eight-man crew and work by himself. Boom. He could not handle a big job I had. He referred me to someone else.

This other guy seemed very nice, and he seemed to know his stuff well. He seemed conscientious. Things began well between us. Then he simply did not show up when he had promised to. I am not used to that happening. I asked what happened. He said he had been given additional work at another job he had been trying to finish up.

It happened again. Both times, he did not notify me. A third time he promised to be there but called to say he could not make it. We finally had a talk. I said I really needed promises kept, so please do not make promises you cannot keep. I said to him that all of us have things come up, so please call me to let me know that, but I was definitely feeling that his promises were not meaning much at all.

He said, "Well, you know, I hate to do it, but sometimes you just have to lie."

I said, "*What?!*"

He said, "Ya, sometimes you just have to tell people what they want to hear, even if you know you can't do it."

This information was not landing well with me. I said to him, "And then what happens the next week when you don't show up, and you have to try to clean up your mess from lying the week before?"

He just shrugged and said something like, "Well, that's just what you have to deal with."

I responded, "Why not tell the truth in the first place? I would much rather be told the truth every time!"

He shrugged. That was the end of our working relationship. Nice guy. We always said hello after that, but his approach was just foreign, unworkable territory for me. I still shake my head over how he saw things, because I know his same patterns continued, and would affect everyone he worked with on into the future.

More on the pain:

So, where can property owners *experience pain?* Well, they can experience pain if there is a breakdown with any of these points. Some points are going to be more important than others, and everyone has their priorities, but any one of them is a potential source of breaking of faith, trust and communication.

When things start to go south, the transition can happen with amazing speed. It can then be extremely tough to rebuild faith and open trusting communication again. Jobs that have gone along with great success can turn sour overnight, perhaps irretrievably.

This is where the importance of good and open communication cannot be emphasized enough. If you have established a foundation of open, honest communication, you are much better able to talk about whatever comes up. You are also much more inclined to give one another the benefit of the doubt and suspend judgment.

Some years ago, I was doing a costly renovation and addition job for a very nice couple who owned three homes. When the excavation machine operator dug down into the ground to prepare for the foundation for one addition onto the house, he found that the soil was all fill that had never compacted. At the level we were supposed to pour the footings, the base of the concrete foundation, the soil was not trustworthy for supporting the foundation as specified.

I called the owner, who was out of state. I described the situation. He asked what I would suggest. I said we wanted to make the footings twice the size and add plenty of steel reinforcing rods. I said I knew it would cost more money than we had figured on, but I wanted to sleep at night and not worry about his master bedroom addition compressing its way into the ground.

He said that he wanted to sleep well, too, so go ahead and do what I proposed. We had more scenarios like this during the job. After a while, the owner said to me "Just go ahead any time something comes up and use your best judgment. My wife and I trust you and know you are looking out for us."

Wow. Not only did that make me feel really good, but I knew that the owner and his wife and I had developed that level of trust and faith together that moves mountains. Their belief in me and the work we were doing also made it easier for me to want to *keep* doing my best all the time. *This mutual trust helped me through what became a highly pressured job due to schedule constraints. Their trust and faith were big positive supports.*

The atmosphere around a job where there is much *mis*trust is palpably different. There is more tension. People tend to keep their mouths closed more, and watch what they do say. It is harder for the crews to come to work. The sourness pervades the air. There is

29

more complaining. There is frequently a lack of cooperation. People tend to protect their own territories.

Frequently a breakdown, and then an escalation of conflict, occur when there is some kind of disagreement followed by poor or no communication. Then everyone is on their own to concoct their own version of what took place and what the others are probably thinking and meaning. Assumptions and conclusions are made. The truth is generally somewhere in the middle, but it may well be lost. It becomes much harder to heal such a rift than to create it and fall into it.

Assumptions of another sort are a common root source of conflict. These assumptions have to do with interpretation. That is why it is so important to state the details in writing and go over them together. I learned this the hard way, too.

Over twenty years ago or so, my former partner Don and I took on a remodeling job that I should have walked away from. However, we were hungry, and the night I went to sign the contract I ignored the obvious signs that it would be better to walk then and there.

In any event, some simple storage was to be added in one corner of the existing kitchen. I wrote "storage" in my proposal. What I had in mind was some shelves. What the husband interpreted was a whole cabinetry built-in. I learned to think and write with the detail of an attorney, in a sense! Not the legal protection and legal-speak, but *details.*

Now I write things such as: "$600 material allowance and $1400 labor allowance for building and painting custom shelving in corner of kitchen. Shelving will be painted with one coat of oil based primer paint and two coats of Benjamin Moore latex finish. Shelves will be constructed of paint-grade ¾" birch plywood with poplar edging . . ." You get the idea. *Written details leave little or no room for interpretation and misunderstanding.*

When I am meeting with customers before signing up a job, even if they are old, trusted customers, I go through the same detailed process. It is good for both of us. I find this process also helps to fix the job better in my mind. And both of us having and reviewing a good written description together helps them too, while it also deepens and clarifies the bond between us. *Most of my customers seem to love the fact that I attend to details like this.* It helps establish and expand our mutual trust and their comfort with me. Many of my customers are detail-oriented people, and I tell them I do not mind at all if they keep an eye on me regarding details, even though they know by the many notes I take that I am very focused on them already.

What are the needs and wishes of contractors?

Again, I don't claim necessarily to have everything covered here, but I think the following points will address a lot of the primary needs and wishes of contractors:

Trust. I dare say that the vast bulk of contractors want to feel that their customers trust them. If there is a lack of trust, or a *perception* of a lack of trust, there is always a cloud hanging over the job that has subtle and not-so-subtle effects. The atmosphere always has some degree of tension. As I alluded to earlier, a solid bond of trust brings out the best in everyone involved. It helps build a positive emotional climate. It naturally invites people to want to help each other and take the best care of each other that they can.

Efforts to establish trust and deepen it are well worth it!

Faith. By faith, I mean that the customers believe that the contractors know what they are doing and will do a good job for them. Of course, it may be natural to have some sort of feeling-out and "well, prove it to me" period at the start. Sometimes this testing period can be skipped altogether, if you establish faith and trust at the beginning. And if there is such a period, hopefully it will be very short, and if the contractor is lucky the property owners will do their best to take a chance and just *have* faith.

"Faith" in this case also includes believing in the contractor enough that the customer does not think they have to watch over the contractor like a hawk at all times. I tell my customers I don't mind if they keep an eye on me, but that is more so that I remember all the details.

If you as a customer do believe that you have hired a competent, honest contractor, I suggest you thank your lucky stars and give them room to do their work.

Respect. You generally know when someone respects you. It is not good to be completely dependent on the respect of others for your own self-image, but when someone does show you respect, it bolsters you and helps you feel good about what you are doing. *Again, respect helps bring out the best in others.*

Timely payments. God, this is a big issue for a contractor!! I hate to say it, but many customers seem to believe that most contractors are so financially successful that they may not even have to work. Many people believe that if they pay a contractor $100, the contractor takes home maybe $85. This is not the case, not even close! If you are reading this and saying to yourself, "Yeah, right," I probably could not say anything at all to convince you otherwise.

There is an old joke, "how do you know when a lawyer is lying?" The answer is "when their lips are moving." (I hope I have not stepped on too many lawyers' toes.) Well, some people believe this about contractors. Some contractors fit the bill here, but there are so many honest, talented ones who do not, just as there are many fine lawyers who are wonderful human beings who live in a spirit of great generosity and service to others.

Contractors have loads of expenses. Payroll, materials, subcontractors, insurance of all sorts, payroll taxes, administrative costs, equipment and vehicle costs, etc., etc. When my ex-partner Don and I first built up a crew in the mid 1980's, I handled most of the job meetings, beating the bushes for work, estimating and contract signings for a couple of years. Having a crew was completely new territory for both of us.

It is said that you cannot describe a headache to someone who has never had one. Well, no one could have told me what it would *feel* like suddenly to realize, "Oh my God! Not only are these people dependent on my partner and me to keep them busy and making money, their *families* are dependent on us for their survival and livelihood too."

This was a sobering realization, to say the least. I would sign up a new house to build, and as I was driving away from the signing I would think, "I have to sign another one. We need two or more going at the same time. This one is only going to keep us going a little while." We now had all these mouths to feed, children to clothe, cars to pay for, vacations they deserved . . . Any employer can understand what I am saying.

As a side note, many trades people who have never been self employed do not get this either, because they have never been there. We had employees who were sure we had money we did not have, even though there were years *that we paid them more than we paid ourselves to survive challenging times.* Don and I can chuckle about this now. But again, no amount of explanation could change the impression.

A related story is that in the 1980's for some time a friend known as Pup worked for us. He had been a partner in a successful dock-building company on Cape Cod. He told me how he showed up at one of his dock building jobs one morning with a new pickup truck, and his guys all had the attitude of, "Oh, yeah, look at him with the new truck. Why does he have all the bucks?" There was resentment and envy.

Pup quietly explained to them that his old truck had lots of miles on it, he got a good deal on this one, and from lots of hard work there was the money available to buy the new one. The resentment and envy didn't budge.

Finally Pup said to *himself,* "Hey, screw it! I work my a** off, I built up this business, I work 60 hours a week, I have to do things the crew never knows about, and if I want to buy a truck I'll freaking well buy a truck! And I don't owe them any explanation!"

Honesty. Tell the truth to the contractor, about what you want, what you expect, whether you really have the money for your project. *This is not a poker game where you try to keep one up on the other guy and fool him. What you want is a win-win here, not a win-lose.*

Clarity and transparency. Related to honesty. Don't hide things. Be as clear as you can be. Don't expect people to read your mind. Ask if they understand you. Is there anything that is not clear?

Communication. A good relationship has to have open and frequent communication. When communication falls off or gets shrunken down as though through the narrow end of a funnel, a great recipe for pain, heartache and disaster starts to brew and ferment.

Ask questions of the contractor. Bring up what's on your mind. Do it respectfully, but do it! Don't assume they know everything. Don't assume they know what your concerns are if you don't bring it up. Don't assume anything at all.

Job satisfaction. Job satisfaction is a high priority to motivated contractors who take pride in their work. For my money, you want to hire such a person. They care even if no one is looking.

I think that this list covers some major needs and wants of contractors.

Now for contractor pain:

As with property owners, contractor pain can occur when any one of the above needs and wishes is unfulfilled. I know that for many contractors some of the _worst_ pain that occurs is regarding money.

When contractors are not paid in a timely fashion, or when legitimate extra expenses are challenged and money withheld as a result, especially when the contractor has already paid out related expenses, _it hurts!_ When they ask for money that was agreed upon and people delay paying them or refuse to pay per the signed agreement, _it hurts!_

Not every contractor has a few hundred thousand dollars of extra money lying around. And even if they do, if money is legitimately owed, please take on the responsibility to pay them in good order.

Having mentioned the issue of extra expenses, that is an arena which _frequently_ is an area of contention and relationship breakdown, and I will explore that in more detail later. What I am talking about here is being aware of the fact that contractors really do have all kinds of expenses that they have to cover _all the time_.

....

Back in the early 1980's I was friendly with a house painter who did some of our work. He told me the story of painting the kitchen of a once famous childhood actor who was a regular in a very popular TV show in the 1950's. The guy was still receiving royalties and apparently living off them.

When Tony finished painting the kitchen the actor said, "You've done a beautiful job. I just wish I could pay you."

Tony laughed, sure the guy was kidding. He was not kidding. They worked out an agreement where the actor paid Tony something like $10 a *month,* and Tony never counted on ever collecting what was owed. Tony had three or four children and lived week to week like many trades people. The customer knew from the outset he did not intend to pay Tony. This story wrenched my gut and made me angry at the actor.

I have heard more than one story of a famous personality taking a "but don't you know who I am?" attitude and expecting to not pay for work, as though it was a privilege to serve them for free.

....

Back in the early 80's my old partner Don and I were hired to build a showcase house high up on a coastal bluff. One of the first things we had to do was re-route an old drainage line from the neighbor's property that ran through the designated septic field of our new customers, on our customers' property.

We talked with the neighbor, and he agreed to pay us for moving the line. We did the work, paid for the expensive pipe and went to collect our money, armed with photos of the work. The proud businessman neighbor took our bill and said, "How do I know what these pictures are? Well, I have your bill. Now it is up to the legal people." Our mouths hung open. We then realized that he had never had any intention of paying us, even though he knew darned well that we did the work and that his old drainage line was illegally running through our customers' property.

We just cannot do things like this to one another! I know that the above neighbor was proud of his business "sharpness" and that he could hoodwink the two young, naïve builders, which he did. But to be *proud* of that? To me that is not something to be proud of. It is sad, and it is part of the matter that fertilizes conflict and ill will.

I do not want to hear the platitude that that's just the way of the world, the way business works. Horse pucky. The way of the world that *I* want to live in is the Golden Rule. Thank God that today ethics are now appearing in business school training, in medical training, and in other fields, and the notion of *being good to one another in every area of life, including business,* is gaining more and more recognition. It's about time!

Of course many, many people already do live exemplary lives in which they treat other with kindness, fairness and generosity. The "greed is good" mentality is starting to die its justified death. Unfortunately, a ton of people are still suffering from the repercussions of that awful mentality.

Many breakdowns in construction relationships either start with or are compounded by poor or nonexistent communication. This is both a source of pain in itself and the opening for more pain.

If things are going sour, if conflict is starting to take precedence over harmony and cooperation, *talk about it!* Many conflicts can be worked out if we will only reach out.

But reaching out is not just about you getting out what you have to say or ask. It is very much about listening, too. It does not cost anything to listen, and quite often simply asking a question or two and then really listening yields way more than we could ever imagine. If you can just hold your anger, frustration, outrage, disbelief, judgments, eagerness to respond—whatever might be bubbling up inside—and *listen*, the needed answer often shows up. It often surprises you, too. You just might hear something that you never dreamed might be the other person's perspective, or willingness to be fair and equitable. There is a wise saying that we have two ears and one mouth for good reason. Take in a good breath and listen.

It may not be easy by any means to practice listening skills, especially with fired up emotions. But it is well worth the effort. If we can ask questions and state clearly and collectedly how <u>we</u>

were affected by something that was done or said, without a tone of resentment or blame, the door is open for real communication and the good chance of resolving whatever is in conflict. It is so easy to decide what was behind someone else's words or actions without checking it out.

Bringing tough situations and tough conversations into the open can defuse things simply by exposing them, too. Just getting things off your chest, hopefully with respect and asking what was behind words and actions instead of assuming, often relieves a lot of steam in the very act of exposure to the light.

On the other side of the coin, MANY disputes start and grow with a *refusal* to listen on the part of someone involved.

There is a marvelous man named David Hoffman. He is a principle of the Boston Law Collaborative. He is a gifted mediator, collaborative attorney, trainer, teacher, and—I am honored to say now—a colleague.

David spoke once at a training session I attended, in which he narrated a powerful example of a terrible result from <u>not</u> listening. A lady business executive was staying at an upscale hotel. She went to the restaurant for dinner. She asked the waiter if a particular dinner had any peanut in it at all, as she was severely allergic to peanuts and could not have <u>any</u>. The waiter assured her the dinner had no peanuts.

She started to eat her dinner and had a severe anaphylactic reaction—her throat was closing up, and she was fighting to breathe and swallow. In fact, the dinner did have peanuts in it. She survived, yet went through a medical ordeal and the terror of losing her ability to breathe and swallow.

The hotel would not even give her an apology. I believe she won a lawsuit of about one million dollars, which she brought simply because of the attitude of the hotel. She said she had to get their attention and not let their grave error go by. She did not need the money. If they had been remorseful and apologetic, if they had asked her what could they do, she would not have brought the lawsuit.

Around the time of hearing this story, I met a doctor who was studying dispute resolution and human relations associated with the world of medicine. He told me that about *seventy-five percent of all medical malpractice suits start because patients did not feel heard by their doctors!* What an incredible figure, and testament to the monumental importance of listening with both ears and a closed mouth.

In my story in the first chapter about the couple with the kitchen dispute, the husband took the big, admirable step of saying he wanted to let go of everything that had been said and done and focus on resolution. He made it clear to the contractor that this was a conscious choice on his part. He could bring a lawsuit, but he did not want to do so. This course of action on his part was textbook ideal practice of dispute resolution. This is what mediators look for. He had the guts to put out the olive branch and to want to step beyond his emotions.

Moreover, what was the final result of his willingness to value successful settlement beyond his own emotions, and his step toward the contractor? His kitchen was completed to his wife's satisfaction in two days' time, and the contractor received his final payment. Meanwhile this dispute had previously boiled on for months with no resolution. You can tell I was very impressed with how this man handled himself. And, his bigness invited the same response from the contractor. I have to praise him, too, because he also responded well.

An attitude of forgiveness and a commitment to cooperate come into play here in a big way.

CHAPTER FOUR

Going Deeper to Improve Construction Relationships

In the last chapter, we looked at the common needs and wishes, as well as some painful experiences, of contractors and property owners. Some of these certainly overlap, as they should. Underneath everything, there are basic human needs and wishes. Priorities vary from individual to individual, and there is no set formula for what will seem important in any given moment. However, we know from models such as Dr. Abraham Maslow's *hierarchy of needs*, after basic requirements for survival such as food and shelter are handled, our preoccupation predictably moves toward satisfying subtler such as self esteem, personal integrity and desires such as job satisfaction—with the ultimate human goal, when every other need has been met, being self-actualization.

These more intangible needs are very important, yet they don't tend to slip naturally into conversations related to construction work. On the contrary, construction projects and the people who carry them out often somehow seem to be relegated to a never-never land that is *unavoidably miserable*. Recall that my customer with the project incorporating a handicapped suite for her aged mother into her second home was told by her friends that she would, *without a doubt*, end up hating her builder. As if "that's just the way life is." *Who says it has to be this way?* To paraphrase marketing maven Jay Wallus, "Who is making up these rules, anyway?"

I would venture to say that people don't commonly associate those in the construction industry with having needs like the ones toward the top of Maslow's hierarchy. Construction projects and everything that goes along with them seem to have to be *endured* rather than enjoyed, appreciated and marveled at.

I would also venture to say for sure that *we do not tend to think of having construction "relationships." But we do!*

The most important underlying factor to me is that it all comes down to human beings and how we relate to one another. It does not matter what the project is, what the size or cost are, who the customer is, who the contractor is . . . It is *people* relating to *people*. To me it is worse than a pity to ignore this fact, because in it lies the foundation for ignoring all stereotypes, and forming and nurturing the human bonds that can utterly transform the process of construction for everyone involved.

I will make some connections here. Both contractors and property owners like respectful treatment and care. They both like to be considered to be acting in good faith. They like to feel good about what they are doing. They both like to have people cooperate with them. They like appreciation. These wants and needs are human nature. Why should they not apply in relation to construction just as they would in any other context?

The big challenge here, as I see it, is seeing what we can do to ensure on a regular basis the best possible experiences and results in relation to construction projects. What behaviors can we engage in? How can we approach things in a repeatable way to give everyone involved the best possible chance to have results they are happy with?

To me, *not* addressing such basic needs and wishes as wanting respect, wanting to feel heard and understood, wanting to feel a sense of cooperation, wanting to have mutual trust and security, and so on, is a major mistake. *Not* addressing these is beginning, and then trying to continue, a construction relationship with two strikes against you and a major thundercloud moving in over home plate. Why not begin with no strikes and clear skies, and everyone happy to be in the game?

In this chapter, we will look at more of what has worked for me in doing this.

The reason I call my business Conscious Cooperation is that I finally realized that cooperation can be a conscious choice. Sometimes it flows out of a great natural sense of camaraderie and mutual enjoyment and respect, and sometimes it takes some work—sometimes a *lot* of work! *The big thing I realized, though, is that you can decide to cooperate to the best of your ability even if you do not particularly like the people you are dealing with.*

This was a huge realization for me.

It is much easier to cooperate with people you really like, people you naturally want to attend to well. You do not have to be guided by your liking or disliking someone, though, in order to give them the best of yourself, be there when they need you and be helpful to them. You can choose to work this way in spite of your personality likes and dislikes.

And the amazing byproduct of such an orientation can be the changing of heart of the person or people you may not have started out particularly liking. As you choose to give your best anyway, and you decide to have an attitude of helpfulness toward them, it is highly likely that a surprising opening will take place in which your world expands. It is likely, too, that the recipient of your cooperative, intentional spirit responds in kind.

Many times I find I suddenly start liking and feeling more compassion for someone who at the beginning I'd really had to stretch myself toward. I have in fact experienced all of this repeatedly, and there has always been a feeling of expansiveness and dropping of barriers washing onto the scene. I have to describe it as the sense of some kind of divine touch. It is a little miracle every time.

I am sure that many readers will know what I am talking about. A good kind of addiction can develop. That certainly holds true for me—whenever I have an experience like this, I always want more!

Here is an illustrative story:

Several years ago I had a lot of work going on. One good project seemed to follow another, and they often overlapped. At the time I had a contract to start the large residential job I described previously, with the poor soil and the big master suite addition. This project had been on my schedule for many months. Meanwhile I took on another large renovation project based on an agreement with the architect that they would cooperate with my schedule and not add a lot of extras to their job, *specifically so I would be available to start the other big job when I was supposed to do so*. I was assured there would not be many changes, and they would be respectful of my schedule and previous commitment.

Well, from about day one, the architect kept altering the new job and adding extra work, regardless of my clear talk about my future work. My schedule breathed down my neck for months. It woke me up during the night. It kept me basically laser-focused on having the phases of the current project planned for and arranged so that progress would continue as smoothly and steadily as possible.

Near Christmas time there was a real crunch period on the job. I had all the subcontractors tightly scheduled. I was on the job most of the time myself, doing labor when I could and just being there to answer questions, make decisions, grab needed materials and help lubricate the workings of this fine time piece that I had created for this phase of the job.

43

A few days before Christmas there was a full house of carpenters, an electrician and a plumber. It was busy and a bit hectic, but the general attitude seemed to be pretty much like Santa's workshop, except with large elves who drove pickup trucks. All of a sudden the muscular plumber *blew.* He started fuming loudly, saying how there were too many people there, how he could not work with such crowding, how he was sick of his job, and how he was going to quit his boss anyway. He did not need this, he said. He did not need the job. He was quitting.

This was my first time subcontracting to this particular plumber, who worked for the largest plumbing and heating company in town. He was a rough kind of guy, pretty sullen with—for me, anyway—a bit of a feeling of potential violence pretty close to the surface. In other words, you did not want to just go over and give the guy a hug.

I stood there dumbfounded for a moment. There went my careful scheduling. The heating guys would be lost to me if the plumber didn't finish ahead of them. We carpenters were trying to stay a step ahead of everyone else so they could come in behind us. All my planning, all my work! At the least I would lose *weeks* if this guy left. And then I would not be able to begin the large job I had committed myself to long before, *which had a difficult completion date as it was!* What to do??

I went over to the plumber and asked him more about what was going on. He continued fuming on as he had begun a couple of moments before. After a minute I said to him, "Let's step off by ourselves so we can talk." I put my hand on his shoulder and steered him off to a vacant part of the house.

First, I let him vent some more. I asked him questions to draw him out. I looked him in the eye and listened as he described how he was under tremendous pressure, how his boss expected miracles from him as a matter of course, and just did not treat him well or appreciate him. Etc., etc.

When he was pretty well spent with his story I told him I could sympathize, and I could feel his pressure. Then I told him where I was at, plain and simple.

"I am counting on your good work. I am depending on you. I need you. If you leave now you throw off my entire schedule, which is already almost impossible. What can I do to help you? I will keep other guys away from you. Can I work with you? Can I patch framing back in behind you so you won't have to do that? Can I drill holes for you? What can I do? I will do anything to help you." I laid it out as directly and as honestly as I could.

I asked if he would consider quitting his boss after my job.

He quieted down but was still breathing hard from his outburst. After a moment, he said he would go out and have a smoke and think for a few minutes. I nodded in understanding, then waited with great anticipation as he had his cigarette. I asked God if this new complication was necessary. God just laughed. I was glad it gave him a chuckle.

The plumber came back in, walking toward me quietly and resolutely. He was clearly different than he was several moments before. He came over to me calmly and in a modulated voice told me that he would finish the job for me, and there would be no more outbursts. He apologized for his explosion. I thanked him from the bottom of my heart and told him that I was serious with the offer to do anything to help. He said thanks, but he would take care of all of his work. I breathed a major sigh of relief and thanked the Big Guy.

The plumber was true to his word. Not only did he finish his work, but it was like a new person had entered his body. He was friendly, professional and cooperative to the end. He even smiled and joked with us during those pressured days. I had witnessed a pretty startling transformation.

My customer saw most of the emotional outburst and the exchange, and was amazed at what had taken place. I was too!

What I had done without even knowing it is one of the main premises of Marshall Rosenberg's principles of *"nonviolent communication."* You ask direct, respectful questions of the person you are in disturbance with. You ask them what they need. You acknowledge what is painful and maybe unfair for them. You show respect for them. And, you clearly state where you are at and what you need. Then you see if you can come together. (I hope I have paraphrased this well, Dr. Rosenberg.)

Before you start thinking I am always just an understanding, laid back kind of guy, make no mistake about it—what I really felt like doing in the moment immediately after the plumber went ballistic was to let him have *my* version of venting! I did not need this crap! Good that I did *not* vent. Good that I was able to be more constructive than that. *I knew that he had to feel heard.* All I did was compassionately draw him out and listen to him for a few minutes, let him know that I too thought his circumstance stunk, then tell him my end of things and ask how we could come together. Whew. It worked. That day is etched deeply into my memory banks.

As I re-read this, I also realize that I treated him like an adult who was capable of having an honest, even-playing-field exchange. The plumber came through from his end with flying colors, and I was most grateful. Something had really shifted in that situation. In what could have become a heated confrontation, you might find that a little miracle takes place. *It is likely that the other person becomes more likable to you, and it is likely that they will not be able to help but act toward you more in keeping with the way that you have chosen to act toward them. The results can be stunning and disarming.*

Your attitude is contagious, and you can choose it—in EVERY moment.

I didn't invent this concept. Many wise people have said it in different ways. It is old wisdom.

It's even nicer working with like-minded people to begin with. Over the years I have put together a group of quality subcontractors who enjoy their work, are respectful and helpful to one another, and typically charm my customers with their professionalism and good cheer. Plumbers help carpenters unload trucks. Carpenters help plumbers carry bathtubs. Electricians and plumbers josh with each other and ask the other to step aside so some IMPORTANT work can get done. We all offer to help one another. There really is a great, cooperative atmosphere that many of my customers comment about. They say things like, "Gosh, everyone is so nice, and so skillful. It is very pleasant to have everyone around." We have a good time together! There is plenty of good-natured joking. We enjoy making one another laugh.

My plumber of several years now (not the same company just mentioned) actually does standup comedy at a local rod and gun club. He is gradually retiring and turning over his business to his son-in-law. The first thing he does when he sees you is immediately start into a "story" (ending in a joke, of course). Most are off color. Most would offend someone. But he is as respectful and cooperative as the day is long, and all the customers love him. He has done very well for himself, and he does not come cheap, but he thanks me for every job and always lets me know if he can do _anything at all_ for me just to call him. He means it. And I offer him back the same.

Several years ago I was newly without an electrician, and needed one quickly. I pulled into my lumberyard one morning, and there was an electrician's truck. I did not know the guy, but he was from my town. I waited until he showed up at his truck. I approached him and said I had never hired any subcontractor on the spot before like this, but would he look at some work for me?

He looked at me kind of sideways and said that he normally did not work for contractors because he had had such bad experiences with them. He preferred to work directly for homeowners and business owners. I told him I didn't blame him, but would he give it a chance?

The short story is that Ted became my electrician, and after a while he started telling people it was love at first sight when we met. He and I would help each other in any way we could. That is such a nice thing to know I can count on, and I am happy to be ready to do the same for him. He is highly skilled and majorly finicky, which suits me well, and he is a real gentleman.

My roofing and siding guys of many years, who are really good, told me that they love working for me because I am so fussy. We, too, help each other in all kinds of ways. Sometimes if there is a crunch I'll pick up a roofing gun or a siding gun and work alongside them, and they help me do carpentry if I am short-handed. It is a great way to work. We often work closely together, as I frequently end up installing or fixing exterior trim that needs to be done before they can take the next step in their work.

I feel very fortunate to have such satisfying construction relationships. These relationships among the workers show through to the customers. I am quick to introduce any worker to my customers too, and praise them to my customers. It is all a group effort, and recognition should be given and shared. I am only as good as the people working with me. Again, old wisdom I am borrowing.

Right around Christmas time at another major renovation/repair job which also was on a tough schedule, I was down on my hands and knees grouting the new kitchen floor on a Sunday morning while my buddy Steve laid granite tile on the countertops. All of a sudden, two shoes showed up in my field of vision. I looked up. It was the homeowner.

He was a soft-spoken attorney with a very wry sense of humor. A good man with a good heart. His wife was an absolute doll. She is an interior designer with taste that I call simple elegance. No flash. Just classy.

In a serious tone and with no smile Fred asked if he could speak with me outside when I had a chance. I said okay. I thought to myself, "What the hell could possibly be going on? We are working our butts off here. It is Sunday, right before Christmas." Truth is, I just didn't know what he might want, and his serious demeanor had me a little nervous. I was bracing myself for the unexpected appearance of that "problem client" I'd heard so much about from my peers, although so far he and his wife and I had done fine together

We went outside. He said to me quietly, "Seeing as the time of year it is . . ." I thought, "Oh, this is starting off okay." Then he said, ". . . and seeing as how hard you guys are working . . ." I thought, "Oh, this actually sounds real good . . ." Then Fred said, "I want you to have these." He handed me four envelopes marked for me and three of the top guys working, including Steve who was there working with me that day.

Then Fred said, "If they aren't enough, just let me know." It was clear then that they were bonus checks.

I said, "Fred, it is incredibly kind of you to even do this at all. It is not up to me how much you want to give. Thank you very much!"

I handed Steve his envelope and said, "Merry Christmas from Fred and Jocelyn." Steve looked at me with wide eyes and a smile and said, "Great! I know just where this will go."

What a great feeling that was. The generosity and recognition from Fred and Jocelyn were much appreciated. We knew that all our fussy hard work was making them happy. When the job was all done, Jocelyn told me how much she loved the expanded and largely rebuilt house, and how pleased she was with the work.

Then she said, "There was only one mistake." My stomach sank a little. After a pause she said, ". . . and I made it." And she smiled. It was a small design issue at the main kitchen window that we had extended out in a bay. Wow, I thought. That is nice praise. They then referred me to their friends, who had the large addition/ renovation job with the bad soil I described earlier. I consider them friends, and they had me return a few times for more work. Their house was on a Christmas house tour a year or two after the job was completed, and it was a lot of fun to visit then and be included in the festivities.

For me this way of working and developing cooperative and often rich relationships in construction is really as much what it is all about as doing quality projects and the satisfaction of getting them done well. This aspect of the whole experience is highly important to me. I guess it is the "gestalt" for me.

This very pleasant example from my work experience also highlights what to me is a basic and tremendously important point: Always look for opportunities to express appreciation for high quality work, high integrity and a pleasant, cooperative attitude!

Later on in this book, I mention my good friend Daniel Stone. Daniel is a highly successful, highly effective organizational consultant who is committed to giving the best of himself to his clients. We have had many conversations about *Conscious Cooperation* and about working with mutual appreciation and respect.

An unexpected bonus occurred recently when I was in the company of Daniel's sister, Katherine Stone. She and I talked extensively about construction projects approached from a collaborative perspective. She was part way through some remodeling work at her own home when we talked, and told me about her wonderful architect and his fine craftsmen. She told me all about the project and showed me pictures. She was obviously very happy about the entire experience, and made it clear that she had openly expressed her pleasure and appreciation to the architect and his team.

Kathy told me she knew that her architect had undercharged her, so she was giving him more money. She was also prepared to tip the craftsmen. I told her she must be wonderful to work for, with that kind of attitude.

She suggested that I make more of a point about expressing appreciation for fine workmanship and superior effort. It is nice to be appreciated and have that appreciation expressed.

I have had the privilege to work for many fine people who have been quick with praise and thanks, and it has been rewarding and heartening. As I said at the beginning of this chapter, we are all human, and sincere thanks and praise can go a long way toward lubricating working relationships as well as personal ones. I make it a point to thank and praise crafts-men and—women, designers, and customers alike.

Kathy Stone also happens to be a law professor at UCLA, who wrote one of the major textbooks on alternative dispute resolution, and has done extensive, well-respected writing on labor and employment law. On the one hand, what a small world! And on the other, her stance of human kindness and acknowledgement of others is exactly *what can help avoid and defuse conflict and make for pleasant construction experiences!* It is so nice to know that for Kathy, with all of her formal study of the literature and her research, this approach of fairness and recognition is not just a theory—it comes from the heart.

I want to suggest that this air of appreciation flows from construction-related professionals to their customers, too. I let people know I appreciate working for them, and that I value their kindness and cooperation. Kindness and acknowledgment given out to others come back to you in return. A reciprocal flow opens

51

up that lasts at least until the end of whatever project you are working on.

So Kathy, thanks for what you said, and thanks for your own great stories.

....

After hearing for years from customers and subcontractors that I ran a different kind of ship than most builders, and had a winning formula that I should somehow try to market, I finally listened to them. I knew that I made committed efforts toward harmony, integrity and devotion to quality.

I also worked at creating a friendly, humorous atmosphere on my jobs. Why *not* enjoy the whole thing? Humor can be disarming, and helps keep everything in perspective. If I can get a customer laughing, I know I usually have a comrade. Humor helps drop barriers quickly and build trust. And as any general contractor will tell you, a happy crew is a real plus! Makes the days move along much faster and more pleasantly.

So one night I sat down and asked myself some questions, trying to slip into the shoes of many customers.

I asked things like:

- What do customers have the right to expect from me?
- What do I have the right to expect from them?
- What do I feel it is my duty to give to my customers?
- What generally makes customers happy?
- Why would I want to hire myself over someone else?
- If I were in the market to hire a contractor or subcontractor, what would be my priorities? What would I look for in this person?

- What could I do consistently to keep increasing my odds of having successful projects, happy customers, a successful reputation and built-in referral system?
- What might make a potential customer apprehensive?

Of course, people are all individuals with varying priorities, but there is a lot of commonality. When I started asking myself these questions and others, I started to get the sense that I was on to something that is given little heed or often ignored altogether. I knew in my gut that I was delving into something very important.

Asking these questions honestly also took me around the corner into some big changes that I could not go back on. Once I tickled the surface of this exploration, I had to look at myself in a way that I had not done before in my career as a builder. Whoops! I had opened a door that would forever more keep opening. I had entered new territory. It was both a little scary and exciting, because I knew things would never be the same again.

At first I wondered if this was all obvious to everyone and I was just pretty dense, and late coming to the game. I started taking varied trainings in working with people, including advanced mediation, coaching, group dynamics, and so on. At these trainings, all attendees were always asked what it is they do and why they came to that particular training. So inevitably the subject of what I was exploring regarding cooperation in construction always came up.

In the early years of exploring what became *Conscious Cooperation,* I was a bit tentative about saying what I was working on. I worried people would say to me basically, "Well, duh, Einstein. *Everyone* knows this stuff!" For sure, I was always the only builder there. One of the trainers was a delightful, senior figure in the world of organizational consulting. His name is Charlie Seashore. He and I had a running joke that I was simply "your typical builder" who as a matter of course would naturally be working on more consciousness and deeper human relations related to the construction industry.

I asked fellow students, who were often consultants, educators, therapists, human relations specialists and the like, "Is what I am exploring and developing here so basic that everyone knows it, and I am just slow?"

The common response was that yes, in a way this whole "field" of consciousness and improved relationships in construction is based on very simple premises, *but in general they were not being put into action. They are not being practiced.* People said to me, "Do this! Please! It is needed!"

These people told me their own construction stories. Some were horror stories, and they said, "God, we could have used this stuff!" Others told me wonderful stories of finding skilled builders who possessed as much skill in human relations as in their craft—builders who were honest, caring, funny and charming. Those guys could teach *me* things, I am sure, because we all have unique gifts and unique ways of being with others. Maybe they could get one or two things from me in return.

It was quite thrilling for me to experience people being very enthusiastic about what I was trying to unwrap from its cocoon. In addition, people at the workshops and conferences I went to were very intelligent, conscious human beings who were typically working in a field focused on some area of human dynamics, so their thoughtful and enthusiastic responses meant all the more to me. In time, I gained more and more confidence in what I was doing. I asked for feedback and suggestions. Sometimes they came unsolicited. Clearly, I had struck a nerve.

So, as a contractor, how do you think you might answer those questions I posed just now?

....

Now I would like to ask some questions of property owners:

- What do you expect from a contractor?
- Are your expectations fair?
- Do you have preconceived pictures and judgments around construction people in general?
- What do you think a contractor has the right to expect from you?
- Do you see construction people on a par with you as a human being?
- What kind of working relationship would you like to have with a contractor, or architect or anyone associated with construction? What is your ideal scenario?
- Do you think your ideal scenario is possible and reasonable?
- Are you eager to get into your project?
- Do you expect to be pleased and to have a good experience?

My questions here for property owners and the questions I asked myself as a contractor do relate back to the questions I posed near the beginning of the book. There is some overlap. After all, this is about *people*, and some of the basic needs and wishes are the same.

And it all comes down to, from the very beginning, trying to clear the emotional air and starting off with the honest intention of being the best human beings you can be with each other. It comes down to dropping your guard enough to experience the other as much as you can—as the human being they are.

Please note though: I am not saying I think people should be naively trusting of everyone.

I am saying that having the relationships that I am talking about here does call for taking *some* risk on your part to operate more from the heart. It takes some daring to expose your emotional self, and to try to find and engage the emotional self in the other to more of a degree than you might previously have dared. Meanwhile, use your personal radar at the same time. Use your good sense.

The fact is, usually this kind of honesty and dropping of the guard stimulates the same type of response in the other person. Our behavior tends to be contagious, like I said before. This works conversely, too. Tight, distrusting behavior can tend to bring forth similar behavior from your counterparty. *It probably will take extra effort to eventually shift out of this learned behavior.*

Years ago, I received a call from a man who owned a second home in my town. He said he needed some windows replaced in his house. He described the work he wanted done.

Then he said, "I have to warn you. I have not had good experience with contractors, so I am not very trusting of them." He said he had been given my name as a reputable contractor, but he was bringing into the mix his previous bad experience. Previous contractors had not followed through and finished the work promised, and had left him in the lurch with several uncompleted projects.

I thanked him for his honesty and said I would look at his project. I told him that his honesty right up front was a good sign, and that I like to be honest and forthright myself. I told him if we did come to an agreement and I did his work, maybe we could change his impressions. He said he hoped so, but he was guarded. I liked the guy because of his directness.

I did do his project. He and his wife were very pleased with the work and the way in which it was done. He paid me happily and said we had gone a long way toward changing his impressions. I was pleased, too! It was definitely a satisfying experience to do a good job and turn around the thinking of someone who had been burned.

As always, I owe gratitude to the people who worked with me and helped me get the work done. I did not do it all myself, and gratitude should be given where it is due. This couple I just mentioned also openly showed how happy they were, and expressed gratitude to everyone working on their house. They were really sweet people, and it became a pleasure to work for them, which we did again and again in coming years. After a while the wife paid the ultimate compliment when she said that she loved her primary home, but her second home was becoming even nicer than that one. They loved using their second home with all the improvements we had made. This was very rewarding, too.

....

I'd like to take a slight detour now and go back to the spirit of forgiveness mentioned at the end of the last chapter. Forgiveness is a major component of most mediation, including what is called "restorative justice." It is an enormously powerful quality, which can disarm dispute and hurt like nothing else. It is said in some spiritual traditions that if we did not sin, God would have created others who sin, so that when they sinned they would ask forgiveness. I find this to be a remarkably compassionate notion. "Sin" here is not used so much in any strict religious sense, I believe, as with the meaning of any thought, word or deed that does any harm or takes us away from the best guidance for life that is inside each of us.

So, how does a spirit of forgiveness apply to construction relationships? Well, conflict happens on a regular basis in construction. Misunderstandings, unexpected complications and surprises are inevitable.

An attitude of forgiveness recognizes that we are all human. We all screw up. We all do things and say things we look back on and wish we had not done or said. An attitude of forgiveness recognizes the positive qualities behind what may be, or at least seem like, much less desirable ones. It recognizes the fallibility of humanity and chooses to overlook it.

Forgiveness is not demanding, judgmental or blaming. It is not superior. Forgiveness is not about holding on to thinking you are right. It is about wanting to go *beyond* being right. It is about wanting peace.

Forgiveness is from the heart. It is expansive, unlike judgment and sticking to our view of things no matter what, which feed narrowness and stricture. I am not saying there are not times to stick to your guns. I am saying that sticking to your guns rigidly and judgmentally constrict the flow of life. It constricts movement and possibility.

This concept is directly applicable to construction, because when there is disagreement, conflict or misunderstanding, active forgiveness goes a *long* way toward smoothing over everything from minor unfortunate incidents to more serious disputes and problems. It is the spirit of, "Hey, we are all human. I can forgive that you made a mistake here or you did something that I had a hard time with. I choose to look beyond it. I choose to see that you are much more than this mistake or this dispute."

Of course, if the one who caused a genuine problem or hurt is remorseful, it makes it much easier to be forgiving! It is when you feel greatly mistreated and the other side does not seem to agree at all, nor want to make amends, that forgiveness can be much more of a challenge. That is where my hat is off in great admiration to the person who can step up and be forgiving, even if they feel deeply wronged or believe that the other person was gravely at fault in some way.

That can take great guts and strength of character.

Another important component of forgiveness and getting along well is *not taking everything personally.* Someone might be snappy because their wife is really ill, or *they* just found out they are ill, or they are struggling with something big they have not said anything about, or they were raised in a family where snappy was the main mode of communication, or . . . I could go on and on. The point is it's good to remember that it might have absolutely nothing to do with you.

As for errors made, or work that is of questionable quality, I'll give someone the benefit of the doubt if they take responsibility. I also found that it works best for me simply to have quality people of high integrity who do great work, from the outset.

A good friend was once framing additions for me on a renovation project. He came to me one day to say he had made a fairly serious goof which he'd caught late. He could make a repair that would not be as good as if it had been done correctly a few steps back, or he could tear apart what he had done and redo it. He asked me to make a decision. I asked him what he would choose to do if I were not there. He looked at me and said, "I'll redo it," and walked away. This is a great craftsman with wonderful integrity.

Going back once more to my case with the kitchen dispute, the homeowner chose to rise above his anger and that feeling that he'd been done wrong by the other guy. He said he wanted to make a positive step toward resolution, a *conscious choice in which he was setting aside everything else.* He was practicing forgiveness, *and it took a committed step to resolve things now,* forgetting about who might be right or wrong. If he and his wife had not agreed to take this step, I do not think that dispute would have come to a cooperative settlement. With his forgiving stance he chose to step beyond what had taken place leading up to the present moment, and commit to the possibility of a resolution that could serve both parties.

What he and his wife were practicing is a big component of *Conscious Cooperation.* It is the *decision—the choice—*to be cooperative, and hopefully forgiving. This positive stance can honestly do miracles. It can lead to forward movement out of a stuck place, to problem solving, and even to rewarding relationships that just were not on the horizon before.

When we make the choice to believe in the best outcomes in any situation, and commit to do our part to the best of our ability, amazing and successful alliances can form. Pleasant surprises can become a common occurrence. Help can materialize suddenly from totally unexpected sources. Brilliant problem solving can appear seemingly out of nowhere. *And the fun factor—from working together in mutual support, commitment to mutual success, and the appreciation of what everyone brings into the mix—can turn the most mundane and even the most onerous tasks into pleasant, rewarding experiences.*

I have had many days in my construction business where there were a bunch of guys working hard at different tasks, we were all in it together, and what we accomplished put a big, if exhausted, smile on our faces at the end of the day. We solved some serious head-scratchers together, we helped one another, we all worked our butts off, and it felt good! Sometimes it was a wild dance, but the energy was great, and the can-do orientation was unstoppable.

Sometimes it can be hard to maintain such a positive and cooperative approach in the midst of the onslaught of life's less fortuitous circumstances, but I have found that doing my best to operate this way, and approach *my whole life* this way, *always* eventually leads to a way through difficult situations. In fact, it seems to eliminate a lot of difficult situations. For me, turning to inner guidance and inspiration really works.

There is a man named Harrison Owen who some years ago found himself, through career circumstances, unexpectedly forming an ingenious way of working with groups. He says that for want of a better name he called it "Open Space Technology." It is a deceptively simple format for bringing together a group of people, of any size, who have some juicy issue in common. This is not about theoretical problem solving or intellectual discussion. There must be a *real* issue that has genuine urgency behind it. Harrison says that the more urgency of time and emotion, and especially when there is desperate need, the better the results tend to be.

He feels so gifted with discovering and actually simplifying this format that he has taught it to thousands of people, myself included. And he encourages others to go out and use it, not owing him a nickel. Harrison feels that the good Lord gave it to him, and he is passing it on. He is also an Episcopal minister by training and calling. But as he says, the whole world became his parish upon making his discovery.

Harrison has been invited to work with experts addressing a number of international conflicts, including the ongoing turmoil in the Middle East. Several years ago he wrote a book entitled *The Practice of Peace.* He says that peace does not just happen. It does not just fall out of the sky. It takes conscious effort, just as "conscious cooperation" takes some effort, and hopefully those efforts are for the most part enjoyable and meaningfully productive.

I know a man native to the poor, war-torn nation of Burundi, Africa. In fact, we met at a conference that was centered around the "Practice of Peace." Harrison Owen was there and presided over some of the proceedings. The man's name is Prosper Ndabishuriye. You can check him out at www.jrmd.org. He spends all his time raising money, then bringing people together in his strife-ridden country to build simple houses for families who have lost their homes to the awful tribal warfare there. He actually gets members of the *opposing tribes*, the Hutus and the Tutsis, to work *side by side* to build the homes. They eat together and even gather to sing together.

Prosper also teaches schoolchildren about peace, love and cooperation. The two warring tribes have carried out some of the worst violence against fellow human beings in modern times. When I met Prosper he told me that most of these tribe members walk around with a branch or a knife or some other weapon, fully ready to kill an opposing tribe member before they get killed themselves. Somehow things there have spiraled downward into such incredible senseless violence and animosity that basically all reason is gone. It is likely many people there could not even tell you quite why the whole conflict started. It seems that at this point they just "know" they are supposed to hate each other.

Certainly love is hard to find in such a seemingly hopeless situation! But Prosper is one man who feels a calling that he has to keep answering and answering. It brought the vast difference between our two worlds into stark relief when one day he called my cell phone from Africa while I was on a roof with a crew on Cape Cod—nail guns flying around, compressors thrumming and guys yelling to one another. God bless modern technology!

More importantly, God bless Prosper and his inspired work. He stays committed to it year after year.

In the next chapter we will look at stepping out of ourselves, and out of the ways we are used to thinking and behaving.

CHAPTER FIVE

Wearing New Shoes

There is an old saying: "Before you criticize someone, you should walk a mile in their shoes." I could not find a reference for the quote, but it is well known. It may have a Native American origin. In any event, I think it is also great advice.

I already entered this territory with questions for contractors about what might and can customers expect of them, and questions for property owners about what contractors can expect of them.

Another important approach here is for each to try to understand the needs, wishes and expectations of the other. We can go back to Mark Silver's book, *The Heart of Business*. I said earlier how Mark teaches business people to try to put themselves into the shoes of their existing and potential clients. What are their needs? Where is their *pain* that you can provide some sort of healing balm for? What is that pain? How do these clients experience their pain?

This is more than a helpful exercise. It can be an absolute revelation that changes your life. You might never see the people you ask this about in the same way again. You will probably end up with a deeper understanding of them as well as more compassion for what they are dealing with.

I mentioned Ken Cloke earlier. Ken is an amazing human being. He is a master mediator whose mantra is "Go for the heart." He has many tear-jerking stories about heart-full, transformative mediation experiences that changed lives and went to places that would have seemed hard to imagine.

I would like to share one of his stories here. It has nothing to do with construction, but I think it relates to the discussion.

One of the things Ken has done a lot of is work with what is known as "restorative justice." Restorative justice can take different forms, but in his case, Ken has worked with it a lot in the court system in California, and particularly with youth. The court brings victims together with offenders.

Two young cousins were throwing rocks onto cars from an overpass. They shattered the windshield of one car. The woman in the car was terrified and shocked, but was able to get her car off the road.

The boys were caught, and a restorative justice session was scheduled in the court. Ken presided. It is typical that the victim gets to speak first. The boys were eight and ten, I believe.

The woman approached the eight-year-old and *blasted* him about how he could have killed her. The boy burst into tears from the fury of her attack. With humans being hard-wired together as we are, she also burst into tears as soon as he did. She suddenly *got:* "Oh my God, this is an eight-year-old boy!"

When things simmered down, Ken asked what she wanted from the boys. She said she wanted them to pay for the windshield. Ken said that they simply did not have money for that, and their broken families did not either. The boys were from a struggling African American family, and the woman was white, and comfortably middle class.

After a while the woman proposed that the boys come to her house on weekends and wash her car as a way of working off the debt. It was agreed. She then started to get nervous about having these unknown black boys come to her house. She confided this concern to Ken and in the end, with Ken's encouragement, she agreed to go ahead with the agreement.

The boys started coming to her house. After a couple of weekends she looked outside and decided to bring them milk and cookies. She enjoyed seeing their faces light up when they spied the treat, and it immediately became a weekly ritual. She softened toward the boys. She was forming a relationship with them in spite of herself. They were doing a good and committed job of honoring their end of the bargain.

Soon their debt was covered, and they were done with their car-washing sessions. The woman began to miss their visits, so she called them and asked if they would like to come back, and that she would be happy to now pay them for work. They agreed, and this arrangement went on for years. They formed quite a friendship.

Eventually the older boy was finishing high school and wanted to go to college, but he had no money with which to afford it. Who do you think paid for his college?

Ken is just *full* of stories like this in which lives are utterly transformed from the heart. As I said, he is a remarkable man with a huge gift.

Over the years I have had the privilege of taking three advanced training workshops from Ken. In the first one, he had the students team up in threes. One student played a role from their own present experience, something which was presenting a tough challenge. The other two students were the mediators.

I was one of the mediators. The woman who played the real life role chose to "be" her boss at work. She had worked for this man for seven years. She said she was a loyal employee, and as his close assistant she probably knew him better than anyone else at work. Yet, he was still kind of an enigma. She said that for the most part he was good at what he did, but that he was so tight and controlled it could be hard to deal with him and sometimes even hard to understand what he really wanted. She said it felt like sometimes he wanted her to read his mind, as if she should magically *know* what he meant all the time without him saying much. She said she felt compassion for him, but it could be quite frustrating working that closely with him. He was a very tightly closed off inside a well-defended container. Even after seven years of working closely with him, in some ways she did not know him at all.

Ken had told us that the level at which he was training us to work was never taught in conventional mediation training. He wanted us to "go for the heart" of the people we were mediating for. He told us that when we were well tuned in to someone, we would frequently find ourselves leaning in toward them, and our voices dropping down to very quiet levels. He said we would then be "in the zone" with the person and feel a unique, special connection.

As my mediation partner in the exercise and I were working with the woman playing her boss, I saw her start to change before our eyes. She had started out with her arms wrapped tightly across her chest, as she said her boss did. Her mouth was compressed, and there was no smile. She told us she knew he had a lot of pain inside.

After a couple of minutes of trying to draw out the "boss," suddenly the "boss" dropped her arms to her sides. Her mouth loosened up. I thought, "Holy crap! Am I really seeing this? Did she do this on purpose, or is this really happening spontaneously?"

Then I became aware that I had leaned in toward her. My voice had dropped. I *felt* very gentle and knew we were "in the zone" together. This is what Ken had spoken of!

When we were done and processing the exercise, I asked the "boss" if her changes had happened spontaneously. She said most definitely, *and for the first time in seven years of working for him, she now really understood her boss after this brief ten minute exercise.* We were all blown away. It was such a profound experience she and I took some free time later on to try to continue her journey into more deeply understanding her boss. This had been far more than a simple roll play exercise.

When you take the time to do your best to step into someone else's shoes, you open a doorway to knowing them in a way that will likely be quite unlike your previous impressions of them. You will likely enter territory you have never visited. There is a very good chance you will come away with knowledge in your heart that you never dreamed of. This type of knowledge can spontaneously and quickly answer many questions, and automatically remove barriers between you, just like my workshop mate experienced with her boss. A ten-minute journey took her where *seven years of working together eight hours a day* had never even begun to take her. How wonderful, and utterly amazing!

A few years ago I got involved in an unsuccessful attempt to mediate a construction dispute. I felt sorry about the outcome. During our process, though, I asked the two guys involved to close their eyes for two minutes and really try to slip into the shoes of the other guy, and to try to feel what the other guy was feeling. The guy who was able to do this very well is a hulking, tough, motivated and very intelligent man.

He squished his eyes shut. You could see him working at genuinely achieving what I had requested. In a couple of minutes he got it. He was able to experience some of what the other man was feeling, and why. We did not come to an agreement, but this man came away with more understanding of where the other guy was at. The other man, who was *extremely* intelligent, nonetheless was not able to get out of his own place of separation in that moment, to likewise journey into the experience of the man he was in the dispute with.

Compassion and empathy are all about *feeling* other people. Feeling, intuitive knowledge is utterly different from book knowledge and preconceived "knowledge," and also from conventional wisdom. It is deeper. It is qualitatively different. If you are stuck in your judgments or general impressions that do not take in the individuality of another person, you will continue to have very limited knowledge of them.

This is where statements like, "Well, all builders are out to make the most money they can, and they will screw you whenever you aren't looking," are woefully unfortunate and untrue. By the same token, statements like, "All customers are just looking for the cheapest price, and they think I am their slave," fall into the same category. I believe this is the root of prejudice: blanket statements that are supposedly broad truths but in fact are narrow, extremely limiting declarations.

It is far easier to believe in and hold on to blanket generalizations. To try to really *feel* into someone else and what is important to them challenges you to step out of where you might be too narrow yourself, however well meaning you may be. It challenges you to face the possibility that maybe your world-views are just that—merely *your* world views, and *not* the deep and ultimate *truth* you want to believe they represent.

This challenge can be very difficult at first. But the outcome can be remarkably broadening and satisfying.

When I started asking myself those all-important questions about how customers might see me, and what they might expect and want of me, at first it was intimidating. I knew I was opening myself to the likelihood that I might have to step in at a deeper level. *I might have to grow up more!* I knew I was opening a doorway to having no choice but to take responsibility for what I found.

In starting to take this challenge, it was clear that I would probably have to change some of my thinking and some of the ways I had been doing things. I would have to give up some of my set conclusions, some of my more limited ways of operating and seeing others. I would have to give up some of my judgments and generalizations. *GULP!* I also began to see that I would have to be more self-disciplined too, and I would have to take on more responsibility.

A friend once said that with inner knowledge comes responsibility. Once you see and understand something deeper, if you don't put your learning into practice, particularly with others, your knowledge can fade away and lose its life. When you *do* put into practice your new understanding, you open the door to receive *more* understanding, which in turn also carries more responsibility. And so the circle continues . . .

I am grateful that I was ready for that challenge. That is how the universe works, I think. I got the urge to take that inner journey when I was ready to take it, and to embrace what might come out of it. I am grateful I did take it! I had the urge because I was not satisfied with my "construction relationships." Conventional wisdom told me to let this urge go, that I should not expect to either give or receive more from construction than producing work for clients and hopefully making a decent living.

Well, I had to take that plunge, and once I started the journey seemed to kind of take on a life of its own at times. My questions, suggestions and sharing became more direct. The more I went down this road, the more confident I got. The more confident I got, the more my clients treated me as if I deserved their confidence and respect, and then the more I also felt I had earned it. The circle kept feeding itself and re-completing itself.

This has become more than a series of exercises for me. It has pretty much become part of my way of life. It helps me to reserve judgment and to listen better to someone else. At times I really have to work at it, but the result is *always* rewarding.

A PRACTICE EXERCISE OF EMPATHY:

So, contractors and property owners, how about trying this exercise?
Here are some suggestions for trying it:

- Sit in a peaceful and quiet place. Get comfortable.
- Take some calm, deep breaths. Let yourself relax.
- Do your best to let the day just slip away.
- Let your mind relax.
- Picture yourself peacefully going inside your heart.
- Choose a person you would like to empathize with.
- When you feel yourself in a peaceful place, picture the person you want to "step into." Get a good image of them.
- Try to *feel* them. Do your best to let down your guard. Try not to think. Just feel and observe. Keep them in your mind's eye and in your heart. Try to avoid any judgments or conclusions.
- Be as open and receptive as you can possibly feel—be an empty vessel. Let the essence of who they are come to you.
- Hold the intention of learning what it is they need and want.
- Ask to be shown: What is important to them? Be slow here.
- With that intention, just quietly wait for what you feel and see.
- Stay in this place until you learn something new, or you understand at a deeper level than you ever have before.

What came up?

To pull this chapter together and add a little more discussion, good construction relationships (and any relationships, really) have a lot to do with a willingness to be open to one another. They have a lot to do with a commitment to do your best and be your best with others at all times. They have a lot to do with attitude.

All spiritual wisdom talks about changing ourselves from within. That is where real growth takes place. It is not about applying the latest pop psychology techniques, although outer practice *can* lead to inner change. Please don't misunderstand me. They can go hand in hand. But the *real* change, the *lasting* change is not just outer behavior. It is change at the core of who we are.

CHAPTER SIX

Commitment

Commitment is another crucial component of a successful construction relationship and a successful project. One of the most common complaints I have heard about failed projects is that at some point the contractor simply left and never came back. They might even have left tools at the job site. They usually did not return phone calls. They simply left. This is a real common and, to those of us in the business, very scary part of the poor general perception of contractors and subcontractors.

Another version of this is that the contractor began with a bang and then tapered off his work progress to a trickle. Communication broke down. Frequently, he did not return phone calls. No explanations were given, or the explanations that *were* provided sounded a lot like excuses.

I have heard property owners say that contractors *never* finish a job. Well, of course lots do, but that impression is so common that many people believe it. Many people believe that they will *always* have to come in at some point and finish what the contractor did not, even if it's just a few small things at the end. Apparently, they have experienced this scenario, or have *heard* this is what they should expect, so it is what they believe. Unfortunately, there is a lot of truth to the awful general reputation, and to this aspect in particular.

Aside from some private cases, I have done volunteer mediation in small claims court for years. Construction disputes are not uncommon. There is always some sort of breakdown of trust. The contractor did not come back. The contractor did not keep appointments. The contractor did not do what they said they would do. The contractor did not honor their contract, if there was one. The contractor showed no respect . . . The stories are painful, and *way* too plentiful. At some point you can always see either a lack of commitment or at the very least a breakdown in the commitment they started out with.

My inner search, continuing to ask myself the types of questions I was reviewing in the last chapter, led me to realize *just how important commitment is in the construction process.* There is the commitment to be the best you can be with the project. There is the commitment to excellence in the project. There is the commitment to see the project through to completion. There is the commitment to serve the customers the best way you possibly can, which includes nurturing the relationship the whole way through. And there is the commitment to honesty and forthrightness.

There is also commitment to honoring yourself and your inner guidance the best you can along the way. In truth, I had a rough road with this one in some ways. Being a "helping" kind of person, I often did not take as good care of myself and honor myself as well as I did others. I am sure plenty of people can identify with this statement. It is quite common, and I am still working on getting better at it. It is related in the Sufi tradition that in order to be merciful with others you first need to be merciful with yourself. That is a deep one, and hard for many people!

Another important facet of this, the ability to pick up on and tune in to the feelings of another, and to their essence and integrity, is an extremely valuable quality when it comes to honoring your own needs. I think often times we do not pay enough attention to our inner guidance system when it gives us a "heads up" about a potential customer or job. Or we might misread it because of past experience, or we simply don't trust it. It is a part of us that needs to be treated gently. It cannot be coerced. It is wonderful to have this quality well developed. I think someone who has this highly developed ability to read others tends to also know themselves well.

I learned to not take on any job that I did not feel good about. If I felt good about the project and the customers, I could feel good about my commitment to finish in quality fashion. I just thought in terms of finished projects and happy customers from the very beginning. This became a powerful force in my orientation toward my work. Then it became simply a matter of how to get there, which steps to take in which order. Yet, the commitment had to be there first. I knew this was placing a responsibility on myself that I had to take seriously, but it was just part of the whole package.

This also meant I had to listen to my inner barometer. If I did not have a good feeling about a potential customer, I did not have a good and full feeling about committing myself to them and their project. A lot of builders and trades people take on any work that comes their way, regardless if red flags are flying all through their field of consciousness regarding the new customers.

My advice is simple and emphatic: *Don't do this!* When a good friend found out this was part of my advice, she said, "Hey, what about Mr. Mediator, Mr. Cooperation? This goes against what you are all about!" My response was it is *not at all* against what I am all about. I just know that if I get warning signals inside and do not feel a good rapport with someone, it is generally not a good exercise for me to force myself into a working relationship with them. You cannot force love either. It is similar. I learned this the hard way, and I don't want to keep re-learning it!

Back to commitment. You know committed people. You can feel them. My good friend Steve, who sadly lost his battle with cancer not long ago, was a fantastic craftsman who radiated confidence and commitment. There was never a doubt that he would complete with excellence anything he took on. Customers loved him because they could feel this, and he always produced. To be honest, he too probably put work first to his detriment at times. Yet, he sure had a loyal following and a stellar reputation.

I think that most customers know when a contractor—or anyone associated with a project—is committed to them and their project. Of course, some customers have such fear and distrust that maybe their radar for picking up commitment on the part of the contractor is pretty clouded over. Some people are so jaded that they think they see failures to deliver and outright dishonesty even in positive situations.

An attitude and orientation to commitment can be felt, just like confidence can be felt. This goes both ways. You can feel a confident, committed contractor or craftsman, and you can feel one who is less sure of himself, has less of a can-do attitude, and less of an inner promise to take real good care of you.

Commitment is something that is perhaps often ignored, on a conscious level especially, in the equation of construction projects and construction relationships. I think it is supremely important, though. To me a project without commitment or a relationship without commitment is like a three-wheeled car. Unless the car was designed with three wheels with one in the rear and two up front or vice versa, it is not going to get down the road very well as a four-wheeled car with a wheel missing from one corner. I for one can't imagine taking on and beginning a job *without* a sense of commitment that I will finish the job to the best of my ability and to the great satisfaction of the customers—to me that would feel like a hollow enterprise.

I know that many contractors, subcontractors and other related personnel think and function in this way. For them, too, commitment is a truly powerful force. Here is a quote from Thomas J. Watson regarding commitment. He was talking about bigger businesses, but the message applies just as well to small businesses:

> ". . . the basic philosophy, spirit and drive of an organization have far more to do with its relative achievements than do technological or economic resources, organizational structure, innovation, and timing. All these things weigh heavily in success. But they are, I think, transcended by how strongly the people in the organization believe in its basic precepts and how faithfully they carry them out."[*]

The message is even more important to the individual. Think about it. What is your commitment to your work? To your companion/lover? To your family? To the success of your own life, whatever that may mean to you? What is most important to you? Whatever that is, whatever list of priorities you may have, without commitment you will make much less progress, and experience much less fulfillment, than you will by bringing to it as full a commitment as you can summon forth from your depths.

It is said that passion for whatever you want to achieve is the most important quality you can direct toward achieving any kind of dream, and the greatest indicator of ultimate success.

Commitment applies in many individual ways in our daily lives. It fires us up, and gives us a framework in which to function, in which to do what matters to us. There is commitment to specific projects and goals, and there is commitment to core beliefs and a sense of purpose. There is commitment to ourselves, to our mates,

[*] Thomas J. Watson, Jr. *A Business and Its Beliefs—The ideas that helped build IBM*

and to others. For most people there is also commitment to some extent to God, or a Higher Power, or Truth; whatever you may call your highest standard and the presence or faith that guides you and inspires your highest aspirations as applied in this life. Even if you do not believe in God, I'll bet you can identify *something* that represents your highest truth, *your* deepest sense of guidance.

Commitment harnesses and stimulates our energy. It focuses us. It helps define and energize meaning for us regarding why we are here and why we do what we do.

Believing deeply in your own inspired level of commitment is priceless. There is no substitute. You cannot fake commitment, just as you cannot fake sincerity or quality. You can fool some people, but the truth eventually makes itself known. The more commitment you have, the farther and more directly you will travel toward your desired goal, whatever it is you are doing.

If you have a contractor who is committed to excellence, committed to his promises, and committed to a high level of service to his customers, adopt the guy! Or woman, as the case may be. Make sure they are happy! People like this are highly self-motivated. They don't need to be watched over every step of the way.

Commitment to yourself is awfully important, too. Commitment to listen to your inner guide, commitment to live as fully as you can and be the *most* you can become, as the unique individual you are—people who rank high on these scales are people I want to be around. They have achieved a lot of self-actualization. They typically have advanced maturity. They tend to take a good look at themselves before they blame others. They have their weak points, sure, but they take a lot of responsibility for themselves. They tend to be quick to own up to things and not make excuses.

This last paragraph relates directly to the personal questions I suggest asking, and the meat of this paragraph goes a step beyond that. To me, it is crucial to address commitment to one's Inner Guide, however you may define that. Most people believe in some sort of Supreme Being. Many people believe in a divine order to the universe, even in the apparent chaos that exists. But I will take an atheist any day who treats people well and has a high degree of personal integrity over someone who puts on a show of being religious or spiritual yet treats others poorly. *To me it does not matter if you go to a place of worship ten times a day, or if you meditate deeply for hours at a time. What is far more important to me is how you treat people and if you are generally as good as your word.*

We all mess up. We all have shortcomings. We all have weaknesses. I am not talking about perfection here, but rather the honest, heart-felt striving for it in terms of how we treat one another and relate to one another. That is where forgiveness comes in, too. If you know someone is doing their best, it is hard to fault them. If they are being sincere it is easy to appreciate them. If they take full responsibility for their actions, what more can you ask of them, or wish for?

We can *feel and see* sincerity, integrity and commitment. These qualities invite in return the same kind of response. As mentioned before, emotions and actions tending to be contagious. Well, this is powerfully true as well for the admirable, desirable qualities and actions that I just talked about. I have been in some situations where I did not particularly like someone, but they needed help, and I just swallowed and said, "What can I do?" I shelved my ego and my preferences, and I dove in. *And every time there was a transforming effect. When I realized we can actually do this, it was a huge revelation and a big step for me.* My little ole' barking ego could just go take a nap for a while as I simply made myself available to that person. Every time there was a softening effect that broke down barriers at least a little. An opening and deepening between us grew. The air was warmer. I felt bigger, thank God. Something happened at a heart level, which is what I always want.

CHAPTER SEVEN

Use Your Personal Radar

We all have an inner personal guidance system. It is a great gift. I mentioned it previously, and now I want to focus on it.

Some of us are more attuned to it than others. Some use it consciously all the time. Some of us get sudden flashes of "knowing," or sudden reminders, or make sudden connections. We all get our own "reads" on people. The fewer preconceptions or judgments we have, the more accurate those reads can be, as I pointed out in Chapter Five. This capacity operates beyond our usual intellectual "knowing," and being open both encourages it and invites it to develop further.

These are inner senses that can have great meaning for us! The proverbial "women's intuition" is no joke. Men have it too, but it's pretty likely most concrete form guys or framers would not want to admit to that. Probably plenty of highly paid executives would not want to admit to it either. It is not typical "guy stuff."

Even conversations among guys after meeting someone for the first time point up the ability: "What did you think of that guy?"

"I wouldn't trust him as far as I could spit him. Just a feeling. You?"

"Same here. He said the right words, but something was fishy."

There is personal radar at work.

For the first several years in my business, I did not always listen very well to my inner voice, particularly when it came to signing up jobs. Somehow I believed that I could (or maybe I thought I "should") make the best of the situation with any customer, and it was not right to turn away work. I touched just a little on this issue in the last chapter. It seemed foolish and ungrateful in the divine plan to turn away work, especially based "just" on my gut feelings. In the early days, when my partner and I were often hungry for work, my eagerness to sign up jobs was certainly understandable. And there were some painful lessons along the way.

As years went by and I became a better builder, and the core of *Conscious Cooperation* was forming in my brain as well as my sense of myself, my confidence level grew. I became a bit more picky and choosy about my customers. I had paid some dues. My reputation began feeding me instead of salesmanship. Thank God!

I got burned really uncomfortably a few times when I took on jobs and customers when my inner guidance system had said, "Run for your life! Get out of here, sawdust-brain!" Ouch. They always hurt. I always suffered for ignoring my instincts, in ways that I really did not have to.

The big turning point came when I had been in business maybe ten or twelve years. My former wife's good friend referred a friend of hers to me for a nice project that involved building an artist's studio over her garage.

The woman was leaving the country for most of the following year. We worked out the details. We had a contract ready to sign. She was leaving soon. And my insides started doing somersaults. I never *connected* with her. There was a complete lack of feeling and link between us. There was distance and coldness. *Now* I know what my insides were saying was, "Watch out, baby!"

One night shortly before she was leaving I woke up around 1 a.m. in turmoil over her and her job. I felt awful! The next night I woke up around 12:30. Same thing, only worse. Next night it was around midnight.

"That's it," I said to myself. "I just can't do this." I told my wife. I said I knew I needed work and income, but I had an OMINOUS feeling about working for this woman, and we had not even signed the contract! I told my wife at the time I wanted to back out before any real damage was done. She was in complete support.

The next day I called the woman and told her I had a terrible feeling about doing the job, and I had to back out. She said to me, "You can't do this to me."

I said, "I can't do this to me," meaning I could not take the job under the circumstance. I told her, "If it feels this bad and we haven't even signed the contract, this is not a good sign at all. I am sorry for the inconvenience to you, but this is what I have to do." I hung up the phone, sighed a huge sigh of relief, checked inside and saw that all the anxiety was gone. Wow, what a relief!

That night I slept much better, *and did get other work.* That was really my turning point for trusting that somehow there would be other work, that there was no need to feel so anxious about earning a living that I would take on projects *and people* who just were not right for me.

This can be awfully hard to do if you are not used to it! I know that many self-employed people in the construction industry, and any other self-employed people for that matter, can understand all sides of this story. I had to white-knuckle it a bit to step into this seeming void.

The important thing is—You are not going to be compatible with all people—no one is! My advice is to listen to that personal radar. Do your best to be open and honest and honorable, but if your gut and all your senses just feel that something is not right, and you just can't get comfortable with someone even after giving all angles of it due consideration, why subject yourself to trying to make chicken soup out of bananas? It ain't gonna work!

At times, this may be a delicate balance to negotiate, because in search of the absolute ideal client it could be easy to say, "Oh, this person does not feel *perfect* to work with, so I won't take the job." You might be too fussy here and back away from what could be a healthy challenge. You have to be your own best judge. The point is simply that at times the best thing to do is say "No thank you." And your intuition will tell you loud and clear which ones are those times, if you're willing to listen!

I had another job verbally agreed upon one time. I had to modify my written proposal due to some changes, but the red flags were adding up, and I chose to remove myself from the picture before we started anything and signed a revised contract. I felt like I would have had to really put myself through hoops in a way that had my gut churning, and that is no way to start a project! Aside from seeing difficulties in dealing with the clients, I felt like we were not getting off on the right foot.

The clients were very surprised and disappointed, yet they respected my honesty and my choice. These were people I liked, too, yet I knew that at least at that time I should not work for them.

I struggled for a day or so over the decision. Once I made it, though and had a good conversation with the clients, I certainly felt clear and relieved.

In the last chapter, I mentioned my customer who told me right up front that he'd had bad experiences with contractors, and wanted to warn me of that. I appreciated his honesty and integrity, and I felt good enough about him to want to try out working for him. We agreed to do some repairs to his Cape Cod house and then see how we both felt about it from there.

The first thing we did was replace some windows. The customer was very pleased with our work and our forthrightness. He paid us quickly and said that we had restored some faith in construction work for him. In time, he called with another project and asked if we wanted to discuss it with him. We did the work. Again he and his wife were very happy with the results and the relationship, and we became their regular remodeling contractor/carpenters.

In this case, the customer and I made a conscious decision to take a chance on each other, and the results were very good. And it highlights the importance of commitment as well, because once we made that agreement to work together, *I was committed to making him and his wife happy. In addition, I was willing to work hard, both on the project itself and on developing our relationship, to do it.*

This principle applies to both contractors and property owners. If either one of you does not really feel at the very least strongly *okay*, I suggest that you not move forward toward a contract. I am sure there are cases where people started off with a lot of distrust and poor feelings and they ended up fine, but I'm betting they are the rare exception to the rule! Why play Russian roulette if you don't have to?

To expand on the points I made earlier:

- How do you *feel* in conversation with the other person?

  o Do they put you at ease?
  o Do you feel trusting?
  o Do you find your defenses melting away?

- Are they someone you would enjoy being around and getting to know more?

  o Do they make good eye contact with you?
  o Do they seem to be comfortable with themselves?
  o Would you trust them in your house day after day if no one from your family were around?

- Do you feel that they have your best interests in mind?

If you have serious doubts about any of these questions, or others that occur to you, *please* take a good look and do not rush into anything. Idealism can really interfere with this intuitive aspect of checking out one another, as can the opposite tendency, toward rigid judgments based on inflexible standards.

Your radar can remain your close friend all along the way once you actually have a project going. For the contractor: *does your customer seem happy? Troubled? Distracted?* Politely check it out if you are so inclined—it will help you further develop your ability to *interpret* the reads you are getting. In addition, most likely your customer will feel appreciative (and perhaps even a little surprised) that you cared enough to ask.

I have found that a simple, straightforward inquiry into their level of satisfaction usually is warmly received and appreciated. "Mr. and Mrs. Homeowner, I just want to check out if you are happy with everything. Is everything fine? Do you have any questions? Is there anything you want to be better informed about?" Such conversations usually build trust and enhance the ease of working together. Such questions frequently disarm people and dispel fears. They deepen the relationship.

Of course, in asking those questions you need to be prepared for the answers, too! My experience is that if you know you are doing a good job and there is clearly strong mutual respect, the questions only improve the human connections. Also, in being asked those questions, the clients realize that you are creating a very sincere and inviting opening for them. You are inviting their honesty. They know that *you know* that they could say *anything* in response, positive or negative, and you're willing to take the risk that it might be negative. This kind of exchange builds trust, and it might take literally one minute.

CHAPTER EIGHT

Go Team

Some property owners want to be totally removed from anything to do with a construction project. They want to hire people to do everything, from the planning phase on, and they just *do not want to be bothered* with any aspect of a project at all. In their minds, that is why they hire various professionals to serve their needs and wishes. That is fine for them. I think this scenario is much more likely to be true in the very high end of the market. Personally, I like it when the owners are quite involved, but if they have a good representative, that can work out fine, too.

I think a viable exception would be having close contact with someone *representing* the customer who has the authority to essentially act as if they are the customer. *But as always, I would want to develop this relationship and base it on honesty, clarity, forthrightness and the commitment to a fine and successful project.*

Designing, planning and carrying out any construction project is a group effort. Why not make it as enjoyable and cooperative as possible?

I tell my customers that I would like them to be on my team, and I want them to think of me as being on their team, with their best interests in mind. It is fine with me if they make sure I am remembering everything, as I mentioned before, because frankly at this stage in the game my head is probably filled at least halfway with sawdust. I appreciate the opportunity to build mutual trust and respect. Most customers seem to greatly appreciate this, too. I let them know that I want to give them the best of me that I can, and I want honest, direct communication between us from the get-go. Usually this conversation is well received, often with surprise.

For property owners: you should know that your contractors and all the people who work with them will do their best work if they have your cooperation and respect, too. They are there to take good care of you, but it should be a two way street. Try putting yourself in their shoes. They usually have a LOT to be responsible for. There is a lot to coordinate. There are innumerable details to stay on top of and keep track of. There may be delays and problems that are out of their hands, and yet they are still responsible to deliver your job on time and to the best of their abilities. Invariably, something gets complicated somehow, yet the contractors are still responsible for delivering the job as agreed.

It sometimes takes a tremendous amount of stick-to-it-iveness and fortitude on the part of a contractor to keep forging ahead amidst seemingly constant challenges. Recently I did work for a close friend, and the list of complications was like something out of a Stephen King novel. In the course of several weeks, the people working with me experienced a mind-boggling array of dire circumstances in their lives: vehicle breakdowns, severe illness of family members, death of a step-father, and severe injury to themselves *at home.*

In fact, as I write this, the job is not yet completed, and right now we are in the midst of an absolutely giant snowstorm which is crippling everything and has brought work to a standstill. And it is the *fourth* storm so far during this job, in an area of the country *that is not supposed to receive much snow.* Tell that to the big snowman in the sky—we have received well over *two feet* from the present storm alone, and it is still snowing! Go figure.

In addition on this project, we have had delivered a number of wrong products that were *ordered correctly*. We have had delivered *custom-ordered* supplies that arrived broken. Etc., etc., etc. What are you going to do? Sometimes your best efforts get tossed to the wind by circumstances. This can be quite frustrating, but you simply have to roll with it and find a way to still make it happen.

I do have a request of property owners, and that is: If something should go awry on your project that really does seem to be out of the contractor's hands, please be as understanding as you can. The contractor does not control the weather. He does not run the businesses from which he orders supplies—and believe me, he probably tries to be really picky about his suppliers! He does not manage the subcontractor businesses. It IS his responsibility to deliver your job even with complications, but if he is clearly doing his best, please take that into consideration if delays or problems do occur.

<u>Many</u> property owners are just great as customers. They are friendly, trusting and appreciative, they do their homework where they need to and they pay promptly. As contractors, I think we should thank our lucky stars for them.

I'm happy to say that many of my customers come under this category. They understand the two-way street with the contractor and his associates. They are respectful people who do not have the attitude that the construction personnel are there just to serve them like indentured servants with no questions asked and no professional courtesy in return.

Yet there ARE some customers who fall into this latter category. They do not respect a contractor and employees and subcontractors. I simply do not want to work for them. There are those who *could* fall into this category, given their wealth and status, but thankfully their natures are far different. Again, I try to live by the Golden

Rule and appreciate the same in return, in this business and life in general.

I have talked about my addition/renovation job with the poor soil. The homeowners could have bought me and spit me out a thousand times over, but on the day that my former wife and I went to make final financial settlement with them when the job was done, the lady of the house insisted on making us lunch. She wanted to make sure we were fed. What a wonderful, thoughtful, generous way to be. I never forgot that.

Sadly, she passed away not long after the job was complete, and she had not had much time to enjoy the finished product. I went to visit her husband some time after her funeral. He greeted me warmly, took me into the new bedroom wing to show me how they had decorated it and told me how much his wife had loved the house and cherished the time that she did have there. It is so nice that I had the opportunity to know that and have a warm time with the husband.

These stories lead me to stress the importance of the sense of team, and appreciation of *everyone* involved. I make it a point to introduce my customers to my subcontractors and the various workers who are around when the customers are present. I simply like the friendliness, and it adds to the sense of openness and trust on the job.

It is much easier to feel connected and involved when you know people by name and have shared at least a little conversation with them. Not to mention, those people doing the work deserve to be treated as the whole human beings they are, whether they are skilled stone masons or plumber's apprentices. Without everyone's work and skill, no project would ever be completed.

On that same job with the poor soil the owners hired an out-of-town architect on their own. I had never met the man before. My experience with architects has been mixed, to be honest. However, this guy I liked immediately. Right off the bat he let me know that he wanted to make use of my experience, and not interfere or try to micro-manage my work. He was a gentleman from beginning to end, and a pleasure to deal with. A real good member of the team, who went out of his way to make it easy for *me* to deal with *him*. That particular project was large, detailed and time-pressured. It called for excellent teamwork all around, and the architect sure played his role well!

It is indisputably a fact that I could not produce my jobs without everyone involved. My subcontractors know I value them. Moreover, the feeling is mutual. This team spirit on our part transmits to the customers, even if we don't say anything about it. They see how we work together. They see and feel the mutual respect. They cannot help but notice, and many of them comment about it.

With this notion of team also comes a sense of shared responsibility, which to me is actually a relief. My customers know that I consider them to have a role to play in the success of the project, whatever their degree of involvement may be. One of my repeat customers gets extremely involved with every project. She is highly detail oriented and spends hours and days and months researching various options and thinking about them. Then we often design projects together. We put our heads together.

She apologizes for taking my time, but I know how comfortable it makes her to work through her process in this way. And she has *always* ended up with just what she wanted. It is never fancy. It is tasteful and functional, and it makes her happy. She is extremely appreciative, too. It is impossible to not want to please her given her wonderful, respectful attitude. She, too, loves whatever work we do and the transformations we have teamed up on.

Over the years, we have changed the home she shares with her husband rather dramatically. Every change has been made for the sake of function and visual beauty, for their aesthetic enjoyment. She never intends to impress anyone. At the same time, they like others to enjoy their house and to gather there. She and her husband love their home, and at this point I feel kind of like a close cousin in the family. That's such a nice bonus added on to the satisfaction of producing the changes they desire and then clearly enjoy.

I like compatibility and harmony. I think probably most people do. There is a part of me that at times thoroughly enjoys the kind of madhouse days where there are loads of things going on and being accomplished with great energy. A steady diet of that formula, though? Nah—I'll generally lean more toward harmonious with not TOO much pressure . . . even though I do seem to respond effectively to the pressure. Not always in pretty fashion, however (it's good to know one's limitations!).

When there is an open team spirit, energy flows. Work can be fun. You can enjoy the challenges and take pleasure *together* in the accomplishments. Solving difficult, head-scratching challenges can be very satisfying.

On the other hand, if there is a lot of disharmony and poor cooperation, the entire atmosphere inevitably feels clouded and heavy. It is harder to get things done. Things seem to go wrong more often. Tempers are closer to the surface. You keep looking at the clock, anticipating going home rather than saying at 5:00, "Hey, where did this day GO?" When it's really bad you might even catch yourself daydreaming about alternate career paths. *"If only I had become a brain surgeon . . ."*

There is much more likely to be conflict that requires some serious renegotiations, or ends up in legal wrangling, when the team is not cooperating from the start.

When there is harmony and enjoyment in working together the days just go more smoothly and quickly; the time just flies by. The team becomes more than the sum of its parts. People are quick to want to help one another. This is how work ought to be!

This team spirit can also be a conscious choice. It is part of *conscious cooperation*. You can *choose* to work in a team fashion.

You can *decide* to be cooperative, helpful and conscious of everyone involved in a project.

I found it helps the spirit and helps the construction relationships to check in periodically with everyone involved. I will ask the subcontractors, "Are you all set? Do you have everything you need? Can I do anything?" They know I want to help them in any way I can, and they give the same back to me. If they are going out they will typically ask me if I need anything. I love it!

Over the years, I learned that if I just went about my own business and did not check in, things were not said that should have been. The atmosphere and communication were more closed. Details or problems on the job were more likely to be missed and balloon into bigger problems, and be all the more frustrating because I knew it all could have been caught in time and quickly resolved if there had just been a little more communication.

I myself felt more closed and frustrated in those situations. It just did not *feel* as good to be a part of those projects. And for good reason, because we were not connecting on all the levels that make things flow smoothly. Even though conventional wisdom would tell me it was "just business, just do my work and go home," I could not shake off the feeling that something wasn't the way it should be.

It also became clear to me over time that communication begat communication. My being clear and open in my communication seemed to invite subcontractors to do the same, which was great, and the whole job benefited. Open communication and an orientation of helpfulness and cheerfulness create a better flow of work and harmony. A little effort in this direction goes a LONG way. And it is something that people notice.

Speaking of cheerfulness, I had a customer who had serious storm damage to his house. It was frustrating dealing with the insurance company regarding settlements. I ended up getting very involved in the insurance negotiations, and the customer was highly appreciative. I had faith that the big insurance corporation would come through when the customer had serious doubts. They did come through, but after we dealt with at least five different company representatives and adjusters and had way too many meetings and phone calls. It called for a lot of determination on my part to carry it through to resolution.

It took weeks and weeks to finally reach agreements and get the repairs underway. I really felt for the customer. He had never had to make an insurance claim before, and he had little tolerance for the delays and insurance-speak. He also was a widower, and I simply wanted to take good care of him and make his life simpler if I could.

Along the way I decided to try to get him to laugh every day. Well, we ended up sharing many laughs together, he was thrilled with the repair work, and then he hired me for more work. I really enjoyed doing his projects and seeing him pleased, and I enjoyed laughing with him!

CHAPTER NINE

Don't Leave it to a Handshake Alone

A good friend of mine is an enormously intelligent guy. He is an activist, a writer, an editor, a scientist and a former Peace Corps volunteer in Africa. This is one bright dude. He loves doing his own Don Quixote thing, tilting at windmills, fighting the establishment. He has become quite the polished public speaker. Years ago, he and I became "partners in crime" engaging in volunteer community activism. We had quite the experience together for several years.

John also loves diving into things. He took a year away from the working world when his house was under construction. He worked side by side with his brother-in-law builder, also my friend. John has quite a few tools, and has a good sense of construction.

He is articulate. He is a leader. And he blew my mind when he told me that my urging people to use detailed written contracts went against actual practice. I said, "Huh? Are you serious?" John told me that probably 80-85% of all construction work on Cape Cod is done without contracts.

Then I said, "Maybe so, and this is exactly part of the problem!" As if to confirm my words, years after his house was built he and his wife built an apartment addition for a beloved elderly friend whose dying husband had asked them to care for his wife until her end. It is really quite the story. The widow lived there for years.

However, in the process of building the addition they got into a serious dispute with one of the principal subcontractors. This man does quality work, but he repeatedly did not honor promises and did not finish his work. John was beside himself with frustration and anger. All rational, conciliatory efforts failed. It got pretty nasty. They did not have a good written contract, so it was all "he said/ he said."

I urge people to use detailed written contracts exactly so they can try to avoid just this sort of conflict. Contracts clarify the job. They serve as a basis of understanding for everyone concerned. They serve as recorded memory when human memory fails or contradicts. They help you get a deeper sense of the job because the details are all written down. *I use them even with my great, long-time customers.*

Here are my recommendations for written contracts:

- Make the writing clear and to the point. Avoid legalese.
- Make the contract about the work to be done, not loads of self-protection and penalties. Commercial construction is a bit of a different animal, but it can still be approached in a cooperative, harmonious fashion. Commercial construction deserves much more discussion at another time.
- Make the contract about *agreement.* The tone should be business-like and positive.
- Be detailed about description of the work and materials! Do not say "shelving in hallway." Say something like, "shelving in hallway will be constructed of ¾" birch veneer, paint-grade plywood with poplar face framing. One coat of oil based primer will be applied, covered by two coats of latex semi-gloss finish paint . . ." Describe doors, windows, framing, insulation . . . *everything.* If the job is extensive enough, I use a separate specification document and refer to it in the contract.

- Include timetables if that is agreed upon. If there is a lot of trust and flexibility in the schedule, you may not need to include this category.
- List an estimate or price for the job. List a payment schedule, one that is fair to both sides. List when payments are due.
- Refer to any dated architectural plans. All key people involved should initial a few principal copies of the plans. *Make sure the dates match!* I learned this the hard way.
- Include insurance certificates and appropriate licenses.
- Agree to mediation as the next step if you run into any dispute that you do not seem able to work out yourselves. But try!
- Address work change orders and extras.
- Cover the bases. Be as specific as you can. Be *clear.*
- Write without making *any* assumptions if you can. That has become one of my mantras: "Don't assume anything!" I would much prefer to ask what I think might be a silly question and get a clear answer than make a misguided stab at an assumption. *And I want to be asked those "silly" questions, too. No question is silly if you really don't know the answer!*

I think this covers pretty well my take on contracts. In years past, I used the typical American Institute of Architects contract, but I quickly came to feel they were too geared toward penalties and legalese, so I started writing my own.

Using detailed written contracts saves an awful lot of heartache. It dispels a lot of the "but I thought . . ." or "what I remember you said . . ." You are really doing yourself a huge favor to write contracts that describe the job in *detail*. It helps everyone, and you level the playing field*, and it actually helps the relationships!*

It is true that some jobs are completed happily on just a handshake, for sure, but I feel that in almost all cases you are better off using a contract as I describe. I have heard some people say, "If I can't do business on a handshake and our word alone, I don't want to do business." If that works for you, great. The record however, in the world of construction, is historically slanted severely against this practice. That is part of the reason for this book!

At the same time, the *intention* should be one of good will and cooperation. That is why I urge writing contracts that emphasize the work itself. *The intention of good will and cooperation sums up a lot of what I am espousing.*

For me, focusing on these qualities, along with focusing on excellence in workmanship, really transformed my business results *and my experience of doing business.* Gradually, direct referrals and repeat business became virtually my only source of projects. I was quite amazed watching this process unfold.

Then, as I observed and studied the components of successful construction relationships and successful projects as I saw them, my actions and personal guidelines became more and more conscious. A cycle formed.

I asked myself what was I giving that seemed to resonate with customers. What I saw was that I was giving high quality work, a lot of service and dependability. I was helping people feel comfortable and well taken care of. This was part of the high level of service. In time a lot of customers and subcontractors both remarked about this aspect of working with me. It was and is most heartening, and the fact is it helps my construction business stay active.

I have to keep walking my talk.

It is my suspicion (supported through observation) that *many* more people would like to work together in such a cooperative and harmonious way. Yet having the kinds of conversations I am talking about can seem uncomfortable, superfluous, or even scary.

As they were for me when I first ventured into them. They can feel risky for many people. Many people are afraid to drop their guard much at all.

There also could be resistance on the part of construction-related professionals, who might consider what I am talking about "touchy-feely" stuff that has nothing at all to do with construction work. My response, again, is that planning and carrying out construction projects, like any other group endeavor, is all about people. It is all about working together in the best way possible to build positive, successful, enjoyable working relationships. If you try it, I can almost guarantee you will like it.

It is likely, too, that many of these relationships will carry over from working relationships into our personal lives. Such carry-over already goes on all the time, and working in the way I am talking about can lead to that happening more frequently and resulting in more profound connections. I know that the more I put myself on the line at a deeper and more open level, and expanded my efforts to be cooperative and helpful, the more my working friendships grew. I came to really value and enjoy my slice of it.

In order to support this intention of mine, I began to realize that my contracts had to reflect the positive intentions I was working with. All writing has a tone to it. We read things differently as individuals, just as we see things differently. Look at the countless experiments with a number of people witnessing the same event, all of whom afterwards stating very different perceptions of what took place (and each certain that *they* had the story right!). But aside from slight individual stylistic variations, writing to convey an intention carries a tone that we can feel.

The tone of the AIA construction contracts noted several pages ago always struck me as too protective and guarded, with too much emphasis on penalties. In the past when I used them it also seemed to me that all the protective focus of the penalty clauses was geared toward the property owners. It just did not feel right to me.

If you approach business with a "We always need lawyers nearby" mentality, as if that is just how business is done, you push aside some of the valuable human connection aspect. To me, you are setting the stage, inviting things to happen that are more likely to necessitate bringing in lawyers, whereas simply with different wording and intention you might avoid all that. All it takes, it seems to me, is an entirely different orientation from the beginning! And the power of that approach has been borne out for me over time since I started writing contracts that made sense to *me* instead of catering to the risk-aversion of attorneys and cynics.

Two of my best customers have been attorneys, although only one of them actively practiced law. His name is Fred, and he is the man who gave my team the unexpected Christmas bonuses. In our first conversation about his project, he let me know that he was not even going to talk with another builder about his project (after his neighbor-friend had told him he "would kill him if he did"). Fred also said his legal colleagues would run him out of town if he did not have a contract with me.

I had no problem with that, and readily agreed. Fred had his then-attorney son-in-law work with me on the contract. His son-in-law and I hit it off immediately. He suggested using the AIA contract form, and we proceeded to cross out entire sections of penalty clauses and unnecessary boilerplate. The resulting contract focused more on the project, and it worked out fine.

A couple of years later Fred and his wife referred the man with the big addition/renovation job with the poor soil to me. That client was also an attorney by training, but his work was in his large family business. I wrote up the kind of contract I have just described, focusing on the work in detail, avoiding all legalese and penalty clauses. He was happy with the contract, and we did great together.

I wrote in there, as I normally do, that we would do our best to resolve any disputes together, and in the event that we could not resolve a dispute we would turn to mediation as the next step. Mediation is a voluntary process in which two disputing parties use the services of a neutral third party to help create an atmosphere in which the two parties can craft an agreement which meets the most important needs of both. Professional mediators participate in everything from construction disputes to divorce settlements and international negotiation.

In a following chapter, I will say more about mediation.

CHAPTER TEN

Money

They say that money is the root of all evil. I am not sure about that, but I *am* sure it plays a big role in a lot of conflict. I also think it deserves special attention regarding the construction business.

If as either a customer or a contractor your primary focus is money, in some ways I think you are in trouble from the very beginning. Money is important, of course. It is important for both customers and the construction professionals involved in any project. *Yet it should not be the primary focus.* All through this book I advise people to be clear and honest. When it comes to money, clarity and honesty are *vitally* important.

Clients want to feel they are getting their money's worth, and they *should* get their money's worth. I would say that most clients are on some kind of budget, and they need to be careful about money.

Money is certainly important for contractors and subcontractors, too. Just as clients want to receive good value in return for their money, contractors want fair compensation—and they should have it, especially those who do a superior job.

So money can be a touchy issue for both "sides." It often has a lot of emotion attached to it.

I normally do not give fixed bids, but I give honest estimates, and I do my best to stay close to my estimate. Estimating is by definition not an exact science. Years ago as a novice I worked for a very talented carpenter/contractor who told me, "After you have made your most careful, honest estimate, double it—because you will *always* run into something you didn't figure on."

I thought this was an extreme approach, but I have learned to add on time beyond that required to do the actual work in the scenario I envision. Just setting up and wrapping up every day adds up to significant time. Trips to the lumberyard or hardware store take time. Unplanned time can wreck an otherwise accurate estimate.

It is advisable to plan ahead as well as possible, but frequently things just happen that you cannot envision ahead of time, especially in remodeling and renovation work. For some reason, too, it does seem common, for contractors to under-figure time. We all seem to need a "fudge factor."

Although I like to work on a "cost-plus" basis, I normally will also give estimates. And again, I do my best to stay close to the estimate. I take my estimates seriously, and make them as exact as I reasonably can. I always hope someone providing a service for me is doing the same, whatever the service is.

If at the end of the job I feel my estimate was significantly too low, I usually bear some of the difference. It is my sense of responsibility and integrity, which maybe go too far sometimes. Some customers have said to me regarding this practice, "Oh don't be silly. You did the work. You worked hard. I am very pleased. Just tell me how much more I owe you." These people are great, believe me! I feel very respected and appreciated by such people.

I have heard workshop leaders who do training at the national level say they advise against working "cost plus," "because most people don't want to pay overhead and profit to a tradesman or contractor." They also advise using the term "contractor's fee" instead of "overhead and profit." They said if you are comfortable working this way, and it works for you, then God bless you. In most cases, though, they advise to give fixed bids, and either the customer takes it or leaves it. They say your money is your business, and you do not have to reveal your working figures to your clients.

Yet I have never really been comfortable working on fixed bids, because the nature of my jobs has varied so much, and unless you are pretty much doing the same thing day and day out, and particularly a specific trade, it is awfully hard to know just how long something will take to do and what unexpected roadblocks you will encounter. After years of experience, I am normally quite confident giving estimates, yet I always say, "this is not a hard and fast bid, but it is honest."

On Cape Cod, my business evolved to where most of my jobs were fairly large and went on for months. In my written proposals, I stated hourly wages, and overhead and profit. With the clientele I was getting, this was not a problem, and I think they appreciated the forthrightness. I explained ahead of time how I charged.

I moved not long ago, though, and one of my new customers challenged me on the contractor's fee added to some extra work. The original work had gone quite close to estimate, which included overhead and profit, and the man was very pleased with the work. There was no question about the costs as we went along, but toward the end of the original work he asked me to agree to a cap on the billing. I agreed to do that for a certain amount of remaining work, and in fact did not make my overhead and profit target with the capped final figure.

I actually think I made a mistake in using the term "contractor's fee" instead of the sometimes touchy "overhead and profit." In any event, he vehemently refused to pay this fee, which in terms of the dollar amount involved was trivial.

He then warned a friend of his who I had also estimated work for that I "would be adding fees to her billing" if I did the work. I talked with her and said definitely not, I had already figured this into my estimate. She then asked, "Well, what was he so upset about?" I said she would have to ask him.

I then explained the conundrum about overhead and profit. I said that some contractors figure it into their hourly wages, and that one way or another, as a viable business, you have to make some overhead and profit. She said she totally understood. She had worked for a technical services company which gave estimates but often ran over budget and simply billed for their hours, plus overhead and profit. She said to me, "I totally understand this, and you have probably told me too much."

If you pay for office work and accounting services, or you do it yourself, any way you cut it, that takes time and time is money, as they say. Do you rent or own office space? As a trades person, and especially a hands-on general contractor, you easily acquire many thousands of dollars of equipment. All these expenses and purchases cannot come out of your hourly wage if you want to be financially viable. I learned this the hard way, too.

On top of covering basic expenses, any business owner is entitled to a reasonable profit in line with industry standards. You are not in business for yourself just for the joy of being your own boss, although that benefit can be very important.

So how do you win as a contractor with such challenges? I guess keep adapting to the territory and circumstances, and be honest with yourself and your customers. And always keep in mind that, as one highly respected friend paraphrased a quote from a great spiritual master, being honest involves knowing how much to tell someone. Some people can handle the whole truth well, and others reach a point of overload after hearing just a part of it.

I almost never match someone else's price on a project, unless that price seems fair to me. I do not say that I will match or beat any price. I steer clear of those who say they will meet or beat anyone else's price.

This subject involves a sometimes-delicate mix of what you feel you are worth, what is fair, and what the market will bear. If you as a contractor do offer a level of service and expertise that is unusually high, I say you deserve a generous wage.

I can understand customers both wanting and generally needing to work within a certain budget. And, on the other hand, I like being compensated for the work I have done, and know I am receiving fair pay for it. As a contractor, it is not a good feeling or a good situation to know that you have underestimated a project, and you still have a commitment to finish. You cannot withstand too many of those situations and remain a viable business.

I think this is where many problems start, and they often spiral down into a grossly deteriorated state of affairs. I think it is frequently the explanation for why a contractor disappears and does not finish a job, or drags it out, or gets sloppy toward the end. It's because he or she knows they are short of money. Communication breaks down, and then both sides are aggrieved.

Money often starts out as, and then becomes even more of, a difficult issue. I don't have all the answers. All I can do is present my observations and suggestions, and my experience. I have known plenty of talented contractors and trades people who frequently came up short on the money end, all the while certainly meaning well at the beginning.

The world of construction can be extremely competitive. When the economy is not doing well and more people are out of work or short of work, there may be a fearful cutting of prices by many contractors and an expectation of discounts by customers. I know stories of contractors giving such low prices that there is no way they can do a project and even cover their basic expenses. Such a self-sabotaging practice cannot be sustained for long.

So many municipal jobs that end up with lawsuits and defective and unfinished work started out with the requirement that the project must be awarded to the low bidder. It does not take very much observation to realize that the lowest bid may have nothing to do with quality and integrity, and indeed may actually be a red flag to expect the opposite of those qualities once the bidder becomes the contractor. No rocket scientists or PhD. economists needed to figure that out!

In any event, if you address money head on and forthrightly from the beginning, you likely will be able to enjoy smooth passage down what can be an otherwise bumpy road. Over the years, almost all my customers have wanted to treat me as fairly as I wanted to treat them. I tell my customers that I have done an honest job putting together my estimate for them, and that I will come as close as I can in the final billing. They in turn are most often very reasonable when unexpected additional expenses arise.

I believe that both customers and contractors should come out with a "win" regarding money. There is an important factor of mutual respect involved.

Many years ago, my former partner and I received a dream phone call. A retiring doctor and his wife had watched us work near their property and decided they wanted us to build their retirement home, on an astounding piece of land they had owned for years. When our planning talk came around to matters of pricing and the type of contract we should have, the doctor said, "I think the only fair system is "cost plus." With a fixed bid someone loses, either you or us, because there is no way to know exactly what the costs will be."

I was quite amazed with this statement and this man's sense of fairness, and I never forgot it. I had to agree with him. I did still put together a diligent estimate for them, and in the end, we came pretty darn close with the final bill. Needless to say, it was a pleasure to work for them. We knew they valued and respected us.

The nature of bidding or estimating depends on the scope of work, too. The more complex a job is, and the more different trades are involved, the more difficult it becomes to make a truly accurate fixed bid. If a builder is building three styles of houses in a development, he or she can pin their costs down quite accurately. They typically make package deals with both suppliers and subcontractors.

Many subcontractors can likewise break down their pricing to a repetitive formula. Flooring and wallboard subcontractors, for example, often bid by the square foot for basic work. The more fixed subcontractor figures a contractor has, the more accurate the overall estimate will be. Materials can also be figured quite closely as long as it is clear what they are. That is why specific choices for materials and fixtures narrow down pricing to a much more knowable quantity.

Again, my experience and observation have shown that the biggest variant in estimating is generally the time it will take to do things. Most contractors and subcontractors improve at this aspect of estimating with more years on the job. There are computer programs for estimating, and a lot of people use them, but I still want to go over the numbers myself and not just rely on someone else's formula. Furthermore, computer programs cannot really take into account some of the unexpected and unique things that crop up.

When the work is custom and there are unknowns, I feel it is fair to work by a cost-plus agreement, wherein all costs are recorded, and then an agreed "contractor fee" is added to the subtotal. Aside from known extras and changes, one way to keep project costs within an anticipated range is to use a "cost plus with limits" contract. The overall project cost is limited to an agreed percentage or hard number beyond original estimate, aside from extras and changes.

Doing repair work when you just don't know what you will find and one thing leads not just to another but to *six* more is really tough to figure, no matter how much experience you have.

Some customers have either experience or a good natural sense that allow them to understand a lot about the construction process, and then again many do not at all understand what is involved. Then there are those who *think* they understand, but don't really. They might see work being done quickly and seemingly easily on TV, not taking into account that filming frequently leaves out important steps and delays, and compresses time frames, for the sake of packaging a sequence for a show.

If a contractor is really dragging along with work and not working hard, that is one thing. If a contractor is working diligently and a project takes longer than a customer thinks it should, the misunderstanding is probably with the customer. If the contractor is careful and meticulous in their work, takes good care of the property and is personable, I say the customer is pretty darn fortunate and should count their blessings.

People work at different speeds, too. Some people move more slowly and methodically. Others are wound up most of the time and seem to carry a small tornado around with them. Then there are those who get a lot done with little fanfare. No two of us are just the same. *But any day of the week I would value a contractor who has high integrity and really tries to think of the good of the customer and how they can give them quality work and a high level of service—even if it takes a little longer. These factors are worth something in themselves.*

Some customers are wonderfully appreciative of work being done and the people doing it, and others are not. Some are wary and distrustful to some degree, and some are highly distrustful! I don't want to work for people like that, because no matter how carefully you craft your contract with such people, sooner or later you're probably going to find yourself having a disagreement involving money. Furthermore—and to me this is very important—it is just not pleasant to work for people who seem sure that you are itching to put one over on them as soon as they lose vigilance for one moment. Even people who have faith in you but act as if they are the indispensable glue holding any project together can stretch your patience and greatly add to your stress load.

Part of my message here relates to your own self-knowledge and how solidly you embody that in your dealings with customers. As a contractor if you are confident in who you are and what you provide, and you really try to take care of your customers, then I think you are worth the high end of pay in whatever market you are in. I greatly value my trusted subcontractors, and I cannot remember the last time I challenged their rates. To me, they deserve good and even generous pay.

If you have never been self employed, have never had to provide for your own health insurance and at least a fair share of retirement funding, have never had to generate your own work and deal with all the myriad details and responsibilities that go with self-employment, have never had to deal with not having sick days and vacation days with pay (and the list goes on and on!), I humbly suggest that your experience has not been the same as that of your self-employed contractor.

Once I did repair work for a new customer for whom I had previously done one very small project. They told me they trusted me and were comfortable with me, which was good. The repair work was an emergency case, and it was impossible to know what I was getting into. Each exploration into damage revealed further damage. In the end they were thrilled with my work and shocked at my bill. All I did was keep a record of the hours worked and the materials purchased, and put a reasonable overhead and profit at the bottom. Their challenge to me over the bill forced me into some deep self-examination and led to significant expansion of this section on money.

What I arrived at clearly in that situation was this: *I did high quality work with no supervision and no waste of time. I treated their property with respect and care. I kept careful records of the work. I kept in close communication with the customer over what was going on, what we kept finding, and how the work was going. I had the customer's best interests in mind.*

Where I was probably remiss, though, was in not discussing money ahead of time. Even though I could not know what was ahead of us with the repairs, and even though the customer told me they *had to* get the repairs done, we should have had some discussion regarding billing. They had actually given me signals early on about their nervousness regarding money, and I wrongly chose to let them go by unaddressed.

I went over this situation and my disturbance with it with a good friend, and I received a quiet internal message during a few moments of silent reflection: *"Be confident in who you are and what you provide. If you give generously of yourself you can expect generous reward in return. Project this image of yourself, and the right people will be drawn to you."*

I then talked about the matter a few days later with a friend who is self employed in other businesses, and before I could finish the story he delivered the exact same message that I had received internally about having to arrive at knowing your own worth, and then standing firm in that valuation with confidence.

The situation described above challenged me at a deep level to take a good look into my heart regarding who I am and what I provide. The answer received applies, for me, to construction, consulting, mediation, or whatever I am doing. I want to give the best of myself, or as an old friend said of his consulting days, he would look in the mirror every morning and remind himself he had to *earn* his fees that day. What a nice way to present yourself to the world.

The fact is, though, I was already getting referrals and repeat customers. The customer in question had come through a direct referral.

Not only is it no fun to work hard and honestly and have someone not want to pay for all you have done, but it also is simply not fair, aside from the financial consequences. If you have a job with a company, imagine if your employer said to you, "We know you have worked hard and done a fine job on this project, but we just are not going to pay you for the last three days' work." How would that feel? I think it would hit you in the gut, and you would probably want to cry, "No fair!"

In the long run, I was grateful for this particular challenge because it really did help me take a closer look at what I offer and provide in every way. When I looked inside regarding this particular job, I concluded that I had indeed provided a high level of service and had definitely billed fairly. I also knew that I needed to include that story in this book, because I am sure a lot of people can relate to it on many different levels.

Underneath this, too, are the deeper matters of knowing yourself and valuing yourself in general. I am not saying overvaluing, and I am not promoting boastfulness, just an honest knowledge of you, with genuine self-respect. We all have value, and to have a sense of that value in the market place is very important.

If a contractor is not honest and not diligent, I don't think they deserve the same respect that an honest, diligent one deserves. I am not defending all contractors because I am one myself. Rather, I am saying that respect must be earned. Respect based on quality work and excellent service is the cornerstone of a fine reputation.

By the same token, I urge contractors to try walking a mile in their customers' shoes.

The fact is, too, we are not all a good match to do business together, and the internal message I received in the story above told me that I am the right guy for *some* people, and that we will find one other.

In this section on money, perhaps in some ways I have raised more questions than given answers. As always, though, being able to understand more fully the "other side" and *their most pressing realities* goes a long way toward having agreement that works for both. Again, honest discussion in this area can be extremely powerful and successful in the long run. It will likely make it clearer how well suited you are to work together.

Another important part of this discussion regards how honest talk ahead of time can unearth information and perceptions that should be on the table. There is an old joke about someone having Mercedes taste and a Chevrolet budget. No offense to Chevys—I drive a Chevy truck myself. Yet this is a vital point, and only honest discussion can shed light.

At the beginning of my career, I did work for someone I knew. I gave them an estimate. They said it was too high. They told me what they thought the work should cost. I asked if they had actually put together costs to come up with the figure. They said no, their figure just felt right to them. I said that with all due respect, I had sat down and added up the cost of needed materials and estimated the labor involved. Nevertheless, they were still sure *they* were right. I felt like I had walked into Alice in Wonderland, and half a Cheshire cat was smiling at me.

A related point is priorities. I have done jobs where, for example, the wife wanted to upgrade the kitchen, but the husband understood that they needed to repair rot and rebuild the chimney. This is not said to disparage the wife. The situation might be analogous to wanting to take a cruise but you suddenly need serious dental work, which re-prioritizes your wallet for you.

I did another big job where the couple wanted to make many changes to their house. They also needed a $20,000 septic system. They graciously understood this and were able to do almost everything they wanted to, but the needed septic system required setting aside some things on their wish list, at least for the time being.

I can understand and relate to disappointment over having to forgo what you might want and like versus what needs to be done. That is just being human! Moreover, not being able to afford what they would *really* like is very disappointing to some people.

Again, money is emotional! Construction projects in general can be emotional, both happily and painfully so. Personally, I happen to get a lot of joy out of seeing customers happy with what I do for them. *I want them to be happy and to enjoy the results of any project!*

The underlying theme here is to have the intention of fairness and value for both property owner and contractor. Money is important, and often emotional, for both, and having a consistently open, honest discussion about it is crucial.

CHAPTER ELEVEN

The Life-Changing Side of Construction

There has been quite a bit of discussion in this book about actually having enjoyable experiences with construction projects. There has been discussion about being aware of what is really important to clients, what their homes and perhaps places of business mean to them. A personal signature is woven into a building. There is history there as well as memories. There is emotion of all sorts. An environment is created that is far more than the physical structure and features.

I was not going to say more about this subject until I started working for a very interesting woman toward the end of writing this book. She works as a psychotherapist. A few years ago, she lost her beloved husband. 2010 was the first time she'd had any work done to her home since losing him. It was the first time she was ready to make any changes.

I knew her already, as we are neighbors and friends. When she began to talk with me about fixing up her house, changing the interior color schemes, changing some of the flooring, replacing doors, and so on, I knew this was a very big step. She had decorated her home with her husband, whose presence and influence were apparent. This woman had many good memories of their life together in that home, along with the more painful memories of her husband's illness and ultimate passing.

She brought in a decorator, a very nice and conscientious woman. The new wall paint in particular is very different from the original. A lot of wallboard repair has been called for, and the new paint shows EVERYTHING, so we took great pains with the repairs.

The reason this chapter was even included is because of my memory of that friend and customer repeatedly saying to me, "I am so HAPPY!" in relation to the work being done. She said she told a male friend that she kept telling me how happy she was. He said, "A guy doesn't want to hear that. A guy wants to hear he is doing a great job." She asked me if this is true. I told her, "I am *thrilled* to hear you are happy, and I know you wouldn't be telling me you are happy unless you were very pleased with the work."

What became so apparent was that her living environment was being transformed, and in ways that really thrilled her. The transformation coincided with her emerging happily into a new phase of her life. One day we went to a tile showroom together. After some consideration, my customer picked out her first choice for floor tile for the main floor powder room. I rejected her first choice, which was bright red, and she went along with me. Ultimately we agreed on a handsome, unusual tile, which she now loves. I made suggestions about a vanity and counter top, and in short order we had agreed on the vanity and granite top and some nice finishing touches as well. Again, she said how happy she was.

Her friends were happy for her, too, and happy to see she was making changes in her home that revealed she had emerged from much of her grief over her husband's passing. She had also prepared for some of the change by working on de-cluttering and sorting through things, which was something she had not been ready to do before.

It struck me how transformative and healing the renovation work was in this case, and how it was not a unique situation. My customer and I talked a lot about the process, and she was actually having a ball choosing products and color schemes and taking what was a big step in her life.

She could now come downstairs every morning and smile at what was for her a new house, and one that warmed her heart. Being the psychotherapist with a spiritual bent that she is, she was attuned to the layers of this transformation in a way that you don't hear about every day, and as an agent of a part of this transformation I was humbled by what it all meant to her.

So often, we hear horror stories and complaints regarding building and remodeling projects. I think it is important to honor and pay attention also (or maybe instead!) to what can be some really positive and uplifting aspects of this work. It is not just about getting jobs done. There are people's lives in the mix here; their histories, their hopes and dreams, their hesitations and joys, their changes. It is emotional, and this can be in a very good way.

Every day on that project, my neighbor-customer seemed happier than the day before. I felt pleased for her, and her happiness made it very nice for me and for Martin, who worked with me, to work in the positive atmosphere she created. One day she gave both Martin and me Starbucks gift cards just to say thanks.

I'd like to share something here that she wrote about her experience of the project:

"My husband died in February, 2007. We were married for 26 and ½ years. Not only were we husband and wife but we worked together virtually every day for 23 years until his illness forced his retirement. During the 3 years he was sick, nothing was done to the house. In the three years since his death, it served as a physical memorial to him, as I began to understand what my new life was going to look like.

Our tastes were very different in terms of interior design. When I finally decided to do the work, I knew that the look of the house would be both very different and mine. The house went from neutral walls and traditional drapes and furniture, to a house with a contemporary feel and whimsical features. Now when I come home at night I sit in my living room and feel delighted by the color on the walls; the floors, which were dark, are now very light wood. Between the floors and the walls there is the change between darkness and light—as if my new life is beginning. As if the years of grief and pain have forged the new me, and subsequently, a new look for my house. It has become a house to live in, not the house of sickness, death and grief.

I began the renovation work very tentatively. Each choice felt overwhelming. Was the paint color right? Did I want to take down a border I was sentimentally attached to? Did I want to paint this wall? Did I want to leave it white? What I found out is what I had learned since my husband's death: I was not alone. Everyone had opinions, everyone wanted to support me in this process, everyone shared my joy of discovering the new me and the new house.

I need to share that Stuart was also a guiding star. Although reluctant at first to ask his opinion, he has been available for me as I needed him. He helped me pick out tile, he helped me pick out granite, he gave an opinion when asked based on his years of experience. We worked so well as a team that there are now easy "buzz words" we can use that bring smiles to our faces. For example, when we first looked at tile, I saw a piece of red tile and told Stuart I wanted a red bathroom. He looked at me in . . . disbelief? Amazement? Incredulity? In other words, No, he didn't think that was a good idea. We worked together to find a different, more spectacular tile. Now all I need to bring a smile to his face is to say 'red tile'.

Each day has brought its own share of genuine happiness and joy. I truly was not expecting that. My friends have noticed the changes in me and welcome them. I was not expecting that. No one has reproached me and told me that Brian would disapprove or not like a choice I made. I was not expecting that. I never knew that tile, or paint or wood could be so transforming. I certainly wasn't expecting that. The change process has proven to be interactive: the house changes and I change in response. I change and the house changes in response. I feel like I am now surfing ahead of an enormous wave, the momentum carrying me.

The only advice that I would have for people contemplating or planning projects is to be aware that your project may develop a life of its own, which may totally surprise you! You are entering a time of change. Change means just that: upheaval . . . new good things, new unexpected things, things you can make different, things you can't make different; good decisions, decisions you would like to remake. Time passes quickly, time moves slowly. Change directs you; the universe is benevolent, serendipitous things happen, you can be flexible or rigid. At the end, you are different."

To add a little more to this discussion, it is striking me hard as I write this how important and powerful it can be if the contractor clearly wants their customers to be happy, and gets real joy seeing the happiness their customers experience, and sometimes the personal metamorphosis that results from the transformative aspect of the work being done on their behalf.

I think it is great to encourage the excitement and joy a customer may be experiencing over a planned or ongoing project. It does not have to be a major project or a top end project to be important to the people having the work done. It is *their* home, *their* haven, *their* place to share with others. That is what counts. This can apply whatever the budget, too.

Of course, people vary a lot in what projects mean to them and for them, and that's precisely why I like to zero in on those questions about "Why would you like this work done, or this house built?" And, "How do you see this project impacting and changing your life (aside from noise and dust!)?"

The project with my neighbor has really brought that process of inquiry to the forefront for me. Because if a customer feels that you the contractor *care* about what your hard work means to them and their life, I think a unique dynamic is added to the project.

Something else comes to mind here. A small construction business is a hard thing to sell. You can sell tools and customer lists, perhaps, but you can't sell the personality and character and skills of the original contractor. Frankly, I have seen ambitious new owners take over a small business with its "customer list," seemingly not realizing that the collection of customers was almost completely built around who the original owner was. *A customer list is really more part of the integral history of the person who built the business in the first place than it is about the business itself.*

The term "customer list" even bothers me a bit. To me it is kind of generic and removed, as if customers can be bought and sold. If the new owner has skill and integrity, and a personality that somewhat matches the founder of the business, that bodes well for retention of some of the original owner's customers. However, I think such a case is much more a rarity than not. To me this is more proof of how important the intangibles are that we have been looking at.

CHAPTER TWELVE

Mediation and Arbitration

Many times the terms "mediation" and "arbitration" are incorrectly used interchangeably. They do both have to do with what is called "alternative dispute resolution," or ADR, but they are very different.

As I just described briefly at the end of the last chapter, in mediation a third party—typically known as a "neutral," and typically having no connection to either of the parties in dispute—tries to help two parties in dispute come together to fashion their own resolution. Both sides get to have their say without interruption. Frequently the mediator has private conversations with each party separately, as well as joint conversations. Hopefully, in the end an agreement is reached, and the two "sides" are the ones who really craft the agreement.

The mediator acts as a facilitator for this process, with the role of helping the two parties come up with the points of agreement themselves. The mediator may ask a series of helpful questions to open up the discussion. If there is a stalemate on some point, the mediator may ask a question like, "Would you like to know what some other people in a similar situation have done?"

There are different personal styles of mediation and a few distinct formats. The typical, original mediation training teaches "nondirective" or "facilitative" mediation in which the mediator tries to be as neutral as possible and does not lead the discussion. They help the two parties have discussion. Sometimes when I am mediating a matter, I tell parties that they are the stars of the show, not me.

In facilitative mediation, the mediator helps clarify and condense points made by both sides. The mediators give quick summaries of each party's points and ask if they have understood the details correctly. Then they point out shared interests and desired outcomes. One of their main objectives is looking for common ground, as well as openings for forgiveness. It is key that both sides have their opportunity to speak and be heard.

As a volunteer court mediator for years, I have seen many instances where some real healing seemed to take place in these facilitative sessions. We have seen many hugs and tears shed. We have seen broken relationships make the beginnings of mending. There have been some really dramatic days at the courthouse! Sometimes we mediators just slump in our chairs after difficult sessions in which we end up feeling we have been present at some divinely altering event. There is a sense of awe at times like those. We feel grateful that we have been given the opportunity to be part of, or even just be present at, such a special shifting of human beings.

"Evaluative" mediation is basically geared toward the facts, which are often in dispute! One person's facts are another person's fantasies. It can be amazing how differently people see things and remember them. Once again, that is why a good detailed description of the work, in a written contract, including how extra work and any unexpected things that occur are to be handled, is so invaluable.

An evaluative mediator points at strengths and weaknesses for both sides, and often describes how each side's case would likely look to a judge. Their involvement with the parties is more active and personal in the sense that they inject their opinions, and they definitely do some steering of the discussion. Their focus is more on legal details and legal strengths and weaknesses than on finding common interests or handling the emotional aspects of the case.

There are generally attorneys involved as an integral part of evaluative mediation. The parties often meet together to some extent, but the orientation is quite different from facilitative mediation. The mediators normally have expertise that relates to the particular case, and are frequently attorneys themselves, as this format is so legally oriented.

Frequently the mediators meet alone with the attorneys. They do a good bit of negotiation and "go-between" activity.

The third principal recognized format of mediation is called "transformative." Transformative mediation is the newest of the formats. The term was coined and described in the book *The Promise of Mediation*, written in 1994 by Folger and Bush. In this format, the parties who brought the matter to mediation really do run the show. The intention is to give full recognition and acknowledgment to both parties, and ideally to help them see and hear each other more fully.

Transformative mediation sessions can get highly emotional and expressive. There is a well-known video of Dr. Bush mediating a session for an extremely charged neighborhood dispute that went way beyond the specific complaint at hand. Picture the scene: the fur is flying, voices raised, tears shed, and Dr. Bush sits there calmly, acknowledging and summarizing everything that's being said. His calm presence has quite an effect in its own right. It is clear that he is accepting whatever emotion and pain is in the room. The intention and frequent result is that by the end of the process all sides feel profoundly heard.

The three formats each have their pros and cons, and in reality I think that many mediators, myself included, make use of both our training and our life experience, *combined with intuition*, in a flowing dynamic.

Going back to my case with the unfinished kitchen in dispute, no mediation training program would ever tell the mediator to strap on his tool belt and work with the carpenters to finish the job and keep the resolution going. That was a hybrid case. I took on the agreed role of judge of work done and assessor of the agreement as well as mediator. From the outset, too, I also agreed to take on a particular carpentry detail that was a big sticking point.

Flying by the seat of my pants, for sure. In the end, we had a happy settlement, even though there were some real tense moments. As a bonus, I also ended up having a good time! That may not be such a surprise, however, because, I was *totally* involved in the process and actually felt that I was in my element.

I felt that I had helped maintain the focus on resolution, and on how all parties could work together toward that intention. I knew when the lady of the house quickly acknowledged she would not expect any extra work not previously agreed to, and when the kitchen contractor agreed to keep his men working, and to have me sign on working in collaboration with them, that the underpinnings of successful resolution were in place.

Perhaps a good quick summary definition of mediation in all its forms is that a third party helps the disputing parties listen to each other, each having the chance to talk and to be heard. The mediator then helps to craft an agreement that somehow works for both. The two parties are integral in any agreement.

"Arbitration" is another animal altogether. In arbitration, a third party is assigned to assess the cases of both sides separately, and then issue a unilateral decision. The arbitrator is not a judge. Many arbitrators are attorneys, but they do not have to be; I know a retired engineer who is a mediator and an arbitrator, specializing in construction disputes.

Both sides agree to the choice of the arbitrator. There is binding and non-binding arbitration. In binding arbitration, it is agreed upfront that the arbitrator's decision is final and will be followed. The decisions of non-binding arbitration do not legally carry the same weight since they do not have to be followed. Following such decisions is voluntary.

In arbitration, as in evaluative mediation, the merits and weaknesses of the cases of both sides are reviewed. However, arbitration often leans much more heavily on the side of legal resolution than mediation typically does. In arbitration I would say the factual details, and in particular the legal details and precedents, are more important. In mediation, and especially in more heart-centered mediation, it is not so much about who is "right," or specific facts or precedents. It is more about mutual forgiveness and what is truly important to the people involved, *underneath* all the details of the dispute.

This underlying component can be the field for enormous, life-altering, expansive change. This is why a trainer I once had for a workshop said that he *loves* conflict: because beneath it all conflict is energy, and energy can be shifted in surprising directions and used toward unexpected change. I really liked how he put this.

So, mediation and arbitration pretty much comprise what has been known as alternative dispute resolution. You can see that the intention is to work out disputes without legal proceedings, which are usually costly and painful in multiple ways!

There is another emerging alternative field here, though, which should also be noted. That is "collaborative law." David Hoffman from the Boston Law Collaborative, whom I mentioned earlier, is a collaborative attorney and mediator, and in both cases a practitioner *par excellence*. He is a sweetheart of a guy, and he works with "opposing" attorneys to come to agreement and resolution. This collaborative law format is frequently used for large cases and where there is a lot of money at stake. In David's case, he IS a collaborative, cooperative guy with a big heart. Every fiber of him wants to work *with* another attorney, not against them.

The attorneys involved start off from the beginning to work together for the best outcome for both sides, as opposed to trying to rip out the heart and soul of the other side and leave them living on the street. *Ooof.* Much, much more desirable.

CHAPTER THIRTEEN

Be Yourself

Over the years, I have attended many training workshops on running a remodeling or construction business. National level trainers seemed to say the same thing; as a construction business owner, you are a white collar professional. Period. Supposedly, that is what customers want to see.

One well-known trainer said to "Get rid of the pickup truck! Or if you must have a pickup truck be sure it is a nice, late model truck. Also put away the tool belt. Customers do not want to pay decent fees to a working guy." Supposedly customers, especially people of higher mid-level and upper level financial status, would only want to pay reasonable, professional fees to someone more like them, with whom they could supposedly better identify. No dirty hands and calluses allowed. Nice tailored shirts and pin stripes preferred. Oh?

One of these trainers said to attend every meeting with a customer dressed in at least a nice sport coat. I tried this a few times and felt more and more like a fish out of water. I felt increasingly disconnected from my jobs, trying to be a manager and salesman only, the "professional" business owner. What I felt was, "Yuck. This just is not me. I am trying to be someone else here."

If it *is* you, great—nothing wrong with it! After a while, however, I took the gamble to be on the job for multi-millionaire owners with my tool belt on and dressed in my work clothes. The truth that came out was that they actually appreciated that my own hands were contributing to their job, and my supposedly to some "suspect lesser brain" still had a lot of working knowledge tucked away. They liked to see me working on their jobs. They liked that they could ask me a technical question, and unless it was in an area of a trade that I just did not know enough about, like perhaps electrical work, I could usually give them a knowledgeable answer. If I could not, I would get an answer from the right people.

The owner of the house with the poor soil for the additions told me that he and his wife much preferred that they were dealing with a hands-on builder. Theirs was the job with the biggest price tag I ever did, and they liked me in jeans and a tee shirt. They liked *me,* not my image, and they valued who I was. And again, this is the business executive who told me he had no idea how I possibly did my job and juggled the crazy management circus that he witnessed firsthand.

So, my advice to both construction people and property owners is *be yourselves*. Be comfortable. For me it came down to being wary of advice that tells you to be someone other than who you are. You just have to decide for yourself what advice is good for you on certain matters, if indeed you need any at all.

Go Forth and Build and Remodel in Harmony

Since I was a young boy I was fascinated with construction and things mechanical. I always had Lincoln Logs, Erector Sets and little tools. I was fascinated with the few tools that my father had. He basically could not stand anything to do with construction or repairs, even though his father had been a builder. I stuck like glue to any workmen who came to the house.

Little did I know that I would end up with my own construction business, which I entered into with eagerness and a lot of naïveté. Little did I know about all the business lessons and challenges that lay ahead. Little did I know how much the human relations end of it all would grab me. Little did I know then just how tremendously important the latter is.

I do hope that this book is helpful. I have done my best to put down on paper my humble "wisdom." No doubt many of you can add to my contributions, and I hope you do.

Construction still fascinates me, and I still work with my hands. I still love appreciating someone else's good work. I still watch cranes swing steel, and framers muscle a frame together and tile specialists do their art. I still love good tools. It is just in the blood. When I told my excavator of many years that I had been trying to get out of the business but it kept drawing me back, he said, "You'll *never* get out of it!" Many guys in the business who know me tell me that construction simply flows too deeply in my blood for me to get away from it. I bleed sawdust.

I think that probably construction and car sales are the two industries in which there is so much dispute and distrust that many people automatically cringe when either one comes up. The general impressions are typically awful. The fact is there *is* a lot of behavior in both industries that deserves distrust and real cautiousness.

Yet, construction projects and the relationships involved *can be highly enjoyable and positive, and also highly successful and efficient if handled well!*

May you go forth, and build and remodel successfully and in harmony. God bless.

CHAPTER FOURTEEN

Exploring the Depths of Conscious Cooperation

I imagine it is clear that the concepts and spirit of *Conscious Cooperation* can actually apply to any interaction between two or more people, and for any purpose. I suppose they even apply to our relationships with ourselves. Underlying everything is the intention to share and contribute the best of yourself that you can, while embracing to the best of your ability the nature, qualities and abilities of others. Oftentimes the outcome can be greater than the sum of the parts.

It's good to always remember the importance of faith in one's daily endeavors. Faith in positive outcomes is hugely important. Faith in possibility is hugely important. Faith in you and in others is tremendously important. If you do not truly believe that you and others can accomplish wonderful things together, and perhaps experience great change and surprising results, you are greatly hampered.

For me and many others in the fields of dispute resolution and various forms of consulting work, faith in the Divine or a "power greater than ourselves," to borrow the phrase from Alcoholics Anonymous, is also invaluable. You do not have to be religious.

You do not have to subscribe to any form of religion at all. You do not have to believe in God, and if you do, great. Yet a faith that there is *something* greater than us that can provide inspiration and healing, and surprising solutions and openings in our daily lives, is a major benefit.

Call it inspiration, wonderful flashes of intuition, divine guidance; it does not matter. I will just say that for me this part of existence is crucial, and I depend on it every day. It can work for resolving disputes, bettering relationships, resolving construction challenges, *and forming the beginnings and the core relationships for a construction project that invites openness to inspiration for everyone involved.*

In my own case, I ask for the Divine or this Power Greater than Myself to infuse any project and the relationships involved, and I turn to it daily along the way. Time and time again I have experienced surprising and unexpected solutions to all sorts of work challenges, and wonderful people showing up in my life, to give two examples. For me actually *looking for* wonderful and even miraculous occurrences has become a fundamental part of the way I live. These occurrences can be seemingly extremely mundane, and I want to appreciate wonder in the mundane.

Recently I took my good customer/neighbor to my tile supplier to meet with their excellent designer to decide on tile for two bathrooms. I was thinking along the way how I needed to talk with my tile installer, who I met through the tile designer.

The supplier had moved to a new location that I had never been to. I was relying on my GPS to get us to the new location in a confusing industrial area. As I looked for the correct complex I saw a pickup truck that looked just like one of the trucks my tile installer Ken owns, and I said that to my customer. Well, it WAS Ken. He had spotted my truck pulling in, he knew I was searching for the right place, and he was actually waving us in.

Ken ended up joining us for the whole meeting with the designer, which was perfect. I consider having run into him no accident and a perfect little miracle that day. What were the odds that he would have pulled into that complex just seconds before I did, and he could join us? I gave internal thanks. By the way, Ken and I also have the same birthday, another rare little event. We get along great.

Underneath the intention of helping to foster cooperation and harmony related to construction is my deeper, more general intention to contribute what I can in my small way to the betterment of the world. I could say "humanity," but I also include other life forms, like pets and wildlife that mean a lot to us. My dog Simba, laid to rest in the summer of 2009, was a fixture in my truck, on jobsites and at lumberyards for sixteen years. He had more friends than I did, four legged and two legged.

I receive inspiration and contribution from different places, and I do believe that you never know who might share some real wisdom. As mentioned earlier, I was once co-mediating a construction dispute in a court setting, and the guy who broke the log jam was the *flooring installer.* He humbly got the manager of the company to see things from a broader perspective, which then quickly led to a fair resolution. The guy was absolutely perfect, and humble about it.

As I thought I was coming to the end of this book I got the strong notion that the wisdom and input of others regarding the subject of *Conscious Cooperation* might be helpful. Also, naturally they would contribute observations and good advice, and likely some good stories beyond my own.

So I interviewed a number of people who I know in one capacity or another. I have a lot of respect for all of them, and they are quite a cross section of humanity, as you will see. This chapter of interviews became the longest one in the book. I interviewed high-level people in the field of mediation and alternative dispute resolution, people in construction and related fields, and customers of mine. I hope it is useful to you.

Some of the discussion and some of my related comments may not be for everyone. Some people may think that not all the discussion has something to do with construction projects. For me the whole matter of approaching one another with an intention that everyone has a mutually successful experience and everyone is well served has a LOT to do with construction.

However, please just take what might be useful to you. I am passionate about this approach to construction and life itself. If some or all of the points here ring true for you, I will be happy.

I have separated the interviews and related discussion into a few categories:

1. Professional mediators
2. Professional people in construction and related positions, and
3. Customers

....

Professional Mediators:

I have mentioned Ken Cloke and David Hoffman, both wonderful mediator/trainer/authors. They both graciously gave me extensive interviews, and were the first ones I interviewed. They both embody the spirit of deep giving, and deep perception of others, in a remarkably nonjudgmental way that encourages unusual resolution of differences and even sudden radical change. Both have great stories from their careers, and contribute a lot to the world by their very natures.

I discussed *Conscious Cooperation* with both men, and asked for their wisdom in any way related to the subject.

Ken Cloke responded with the following: ". . . I just finished writing a book which came out last year, and the title is *Conflict Revolution*, and the subtitle is 'Mediating Evil, War, Injustice and Terrorism' . . . The idea of the book is to take what we have learned in conflict resolution and begin to apply it more broadly to other aspects of our life. I think that is what you are trying to do in the questions you have been asking and the work you are trying to do, which is clear that part of what mediation consists of is conflict resolution, with the idea of taking specific conflicts and trying to work our way through them.

"But a second and deeper thing to think of is a set of procedures and an attitude toward relationships that is not limited to a particular conflict, but actually is potentially applicable to all sorts of situations, including the chronic sources of economic and political conflict. I know that's not your specific focus in this, but it connects to it, I think, because what I was trying to do in the book and what I think you are trying to do in your work is to apply conflict resolution to a lot of things that it has not really been applied to before. In doing that, what I've realized in the process is that there is something really remarkable that's hidden within this technique that has a tremendous capability of transforming the attitudes and intentions as you are describing them; of people who are in relationship with one another."

Ken is one of the brightest, most perceptive and deeply compassionate human beings I know. He is charmingly disarming. He has worked hard on his inner self to be able to "hold" conflict for others, while at the same time holding a huge possibility for radical change. He has so many stories where huge change occurred, and powerful conflict melted away, through people being able to drop their rigid defenses and combative stances and listen more with their hearts. Ken is truly a master at helping this to happen, and he fully understands that conflict is, in fact, energy, and that energy can move in different directions. He also understands that *any* conflict is an opportunity to go deeper.

Conflict is actually the opportunity for deeply positive change, and it is frequently a first step toward profound transformation that ends up making us and our lives richer.

This involves listening with more than your ears, and listening to more than words. *Listening with your heart* is really what is called for. It takes practice, and it may seem scary at first, but it reveals information and truth at an utterly different level. It represents a quantum difference from listening for the "facts." The facts and the information have their place, for sure, but where we really *live* is a different place. Moreover, sometimes focus on the "facts" interferes with the deepening and expansion of human connection. Certainly it often gets in the way of conflict resolution.

In this kind of listening, questions arise such as, "What is really important to them?" "What do they really want and need?" "What is their pain?" "What is their joy?" *It is often the unspoken clues that are the most valuable. For example,* arms tightly holding the chest, the down-turned head, the overly soft voice, sudden agitation, deep sighing, or—on the other side of the coin—a sudden light in the eyes, a spontaneous smile and nod of the head . . . and you might simply, suddenly really *get* another person in a deep and humbling way.

I do not mean "get them" like, "Oh, yeah, I got their number!" What I mean is to perceive them at a profound, compassionate and open level that holds no judgment, that naturally wants the best for them *and feels them.* This is when you feel barriers drop away. The importance of differences falls away. In moments like these we realize at a deeper level how connected we all are.

At one of Ken's workshops he opened a bigger doorway for me about being aware of such signals.

Recalling the story I repeated of Ken's restorative justice mediation with the boys who broke the woman's windshield, whoever would have thought that out of the initial violent occurrence that this remarkable relationship would form between the boys and the woman? Well, maybe Ken would have thought it, because he has been part of so much remarkable change.

This is the intention and the energy behind *Conscious Cooperation*. *Conscious Cooperation* is about having healthy, mutually successful construction projects. Nevertheless, at a deeper level it is about the potential to apply these principles to all of life and then take it to places you might never have dreamed of, in your own unique way.

I told Ken how some years ago I started to warn new customers about the stresses that could well show up for them during a project, no matter how well everything was going. I told the example earlier in the book of the customer who told me how about four months into building their house they hit a wall of stress and frustration. And then she remembered my warning her that would happen, and my guarantee that we would all get through it. She remembered, told her husband, and the stress disappeared.

In that phone conversation with Ken I told him, "I was so glad I brought it up . . . It was the first time I brought it up, Ken, and then it became part of my repertoire, too, to warn people about the stresses that might be involved with construction. So it comes down to just really trying to be aware of the human beings involved."

Ken responded, "I think that's exactly it. I think you put your finger right where *it* is. That quality of awareness you were talking about before, about attitude and intention. There is a wonderful little quotation from William James, and it is in the "Crossroads" book. It is in the beginning of one of the chapters where he says, 'The greatest discovery of my generation is that human beings can alter their lives by altering their attitude (of) mind.' I think that is absolutely right. Just altering your attitude towards another person shifts the conflict and the potential for conflict in an enormously significant way. Really makes it kind of gauche to have conflict.

Just not really very polite, and it's not really done. I think that is kind of what happens in mediations, where we just create context that is filled with an attitude of sort of constructive positivity and friendliness.

"I did a mediation the other day for two women who are on the board of trustees for a local college, and they have been fighting with each other and nasty and—you know—really argumentative with each other. Just by doing this session and asking a couple of questions, a big shift occurred. You dedicate your life to this college as they both had; and it's not as if they did not have real problems, but what they discovered was a large part of what had happened was really miscommunication and mistaken ideas about what the other meant or intended."

"And then these two women who were notoriously antagonistic got up at the end and hugged and kissed. It was just really sweet. And the more that happens to you, the more you realize this is what this stuff's about, *and we have been playing on the surface.*"

I put the last phrase above into italics. It speaks so eloquently and so simply to what I have put forth in this book: *that we cannot improve our harmony and our relationships of all sorts, including construction relationships, by sticking to conventional wisdom, status quo, assumptions, judgments and fears.* We have to dare to go into the heart more, dare to listen more deeply, dare to ask seemingly difficult questions and address red flags that spring up, *and dare to care that the "other side" is also well served.*

We have to dare to drop as much of our image as we can. *We have to dare to discover new levels of ourselves and others, which can be enormously rewarding!* We have to dare to look at ourselves more, which can be both wonderful and at times somewhat scary!

Furthermore, just as I found that in response to this deeper awareness I then had to dare to be more responsible and accountable, you may discover the same. For me that doorway led to another, which in turn led to another . . . Maybe this is simply growing up in a sense, but as I followed this path where it led me, I kept feeling bigger and better and more confident at what I was doing. *I also knew I was serving my customers well, and they let me know it.*

This all gets easier the more we practice it. *And, there is nothing to lose!* If we listen openly and with our "heart ears," we perceive new things and get surprised repeatedly, in good ways. We find many instances of feeling as if we are meeting someone in a whole new way.

I also found I could never slip back in reverse through a new doorway I had entered and say, "Whoops, I really don't want to be more aware and offer more service. I want to go back and shut down some of myself."

How far along this line we are each comfortable is very personal too, as is how we "do" it. No two people are alike. There is no formula that fits everyone.

At one of the trainings I took from Ken Cloke some years ago, there was a woman present who was very bright and held a responsible and pressured job. She and I teamed up for one long exercise together, which we both really enjoyed. At one point after a few days, though, she got extremely feisty, challenging Ken in front of the class about something that she had taken wrong. She was quite confrontational. I was surprised, and I felt protective of Ken as I felt that the woman was out of line. People had come from around the world to take this training, and she was basically blasting Ken in front of everyone.

I was more than curious to see how he would handle the situation. Well, Ken lived his message. He very quietly but firmly stood his ground and explained himself, while setting a well defined boundary with her. He never attacked her, though. He made it about her interpretation of him, and how she was holding things, not about her. He showed her great respect while being firm. I was impressed, to say the least. This was a strong, confident woman who was not shy about speaking her mind!

The next day the woman apologized to Ken in front of the whole class, which I also admired.

In our conversation for this book Ken and I also spoke of how *everyone* has these deeper capacities within. In Ken's words:

"One of the things I think is so useful about this experience for me is that it reveals that really everybody can do this. It's not something where there is only a little league of people who know exactly how to do it. It really is located directly in the human heart, and therefore is something that everybody has access to, and I love that. I think that it's really profound."

Is there a downside to what I would say is this perhaps more daring approach to the relationships in construction projects, or in anything, for that matter? Well, you have to be prepared for what might walk through the door once you open it. You might be surprised or challenged in a new way, but the overall experience is very rewarding.

Perhaps another recommendation for this way of doing things is that I can honestly say that I have never once regretted taking this approach, and I have always ended up feeling like something rewarding and larger has taken place. Life keeps teaching.

....

David Hoffman is another fantastically bright, compassionate man who lives the principles and spirit that mediation represents. He and Ken are cut from a similar, yet differently textured, cloth. As the founder of the Boston Law Collaborative, he is a marvelous collaborative attorney, mediator, trainer, writer and teacher. In his spare time, he teaches at Harvard Law School. He is well known as being highly talented at what he does, and a really likable human being.

He also used to be a professional woodworker and owned a shop, which was a pretty mind-blowing discovery for me. I definitely wanted David's wisdom for this book, and he quickly dove in to offer some of it:

"So I think that just as a Zen Buddhist would say that, 'Our pain is inevitable, but suffering is a choice,' I think that in construction work one could say that, 'Disagreements are inevitable, but conflict is a choice.' And how people work through disagreements can make all the difference not only to the financial success of a project, but also to how people feel about the project."

This is not run-of-the-mill thinking or "conventional wisdom." These statements capture the very heart of the matter, and with great simplicity.

David spoke of how important a project can be to a property owner, and how important the successful completion can be to them in terms of satisfaction and finances and how their new structure can affect their lives. "It's going to be real experience they have to deal with," he said. Then, he added, "I think with the owners, sometimes what they may not know is that for the builders, and all the contractors, the successful completion of a project and the satisfied client or customer is enormously gratifying as well."

This is a great point that I think is rarely stated. Think about it, though: don't most people, whatever their work may be, want to have a deep sense of satisfaction from a job well done, including happy customers? Don't *you* feel satisfaction, and spiritually expansive, from doing things well, giving the best of yourself and seeing a positive response from the beneficiary of your work? Of course, we are all different, and some people may seem impossible to please! But by and large I think most of us would agree this sort of outcome always enhances our experience of work, and of life in general.

David continued:

"So there are intangible, non-monetizable benefits on both sides of the transaction, and setting up a relationship, establishing relationship and good postures, those intangible or non-monetizable values seem to me to be worthy goals. When people talk about what the underlying interests are, the interests are not exclusively financial. We all know that reputation means a lot to the people in the construction field, but there are people who do this kind of work who want to feel that they are adding value. They want to feel that they can satisfy the customer because that is satisfying *in and of itself*, not just from the business standpoint."

I asked David what he feels are key components of building crucial trust. He gave a beautifully clear and illustrative answer:

"I think that trust is the result of relationship. Relationship is the bridge only when the trust is built, and trust is one of the foundation stones for successful completion of a project. So before people are going to trust, I think they have to know the other person and know something about them. So, having some opportunity for personal contact and connection is always valuable. When we begin a collaborative law case, one of the rituals we try to engage in is having coffee with the lawyer on the other side, and just sitting and having a cup of coffee and getting to know about the other person helps to create relationship.

"And often I ask the people to tell me about their past that led them to where they are right now. Sometimes they choose to tell me about the past from a very personal standpoint, or from a professional standpoint, or some of each. But either way, just by getting to know the person, we find that we have certain things in common. And I think we all know from experience that when we discover that we have things in common with another person, it helps us feel connected to the other person.

"Another very important way [to build trust] is to be highly responsive and responsible, so that from the very first stage of the project we do what we've said we're going to. We communicate clearly about what our intentions and expectations are so that we don't get a relationship off on a sour note, because sometimes it's very hard to recover from that."

I think David's point here is huge, and reflects what I related earlier that my attorney/mediator friend Jeff Oppenheim also said. That is—and I may be paraphrasing slightly—"When things go south they can go south quickly, and then be very difficult to bring back."

David also spoke further of the need to work at relationships. Using "rituals," as he says—like playing golf with a business associate, opposing attorney, community leader, or customer—is a common way to share enjoyment (at least, if the golf gods are smiling on you!) and to get to know each other in a social context.

David made the statement, "I think the important thing is to understand that if you don't already have that connection, whether you're considering or in the middle of a big project, possibly think about this as a financial aspect, that one has to be conscious and conscientious about the relationship side of the transaction."

David went on to describe his observations when he worked at a big law firm.

"Among the most successful of the rain-makers and client-retainers were the people who knew how to have successful relationships with clients. So the people who returned the client's calls right away, who had a light touch, who had a kind of down-to-earth approach—and those people had these strong relationship skills—were not only very successful in drawing in clients but were also very successful in court, because they knew how to relate to the lawyers on the other side.

"One of the people who fit that description beautifully is this guy Dick Grantham, who became president of the Boston Bar Association. I think he grew up in the Boston area, went to public high school, and went to college locally. He's a wonderful and very bright, down to earth guy, and lawyers who were on the other side of his cases almost invariably said, 'He's a terrific guy.' And Dick was also frequently hired by the parties whom he had opposed in court previously, because they said, 'That's the kind of guy I want on my side next time I have a battle.'

"So I have mentioned those, because I think these kinds of relationship skills that are important in terms of working with the customers or clients also pay off in so many other ways. They pay off in relationships with employees, with suppliers, vendors, public officials and all the people in the universe of contacts and connections that people have to deal with in the construction world. So this is a valuable training you are offering, and not just on how to get the job done successfully and have a satisfied client, but how to have a successful business and a successful life."

In this last paragraph, David beautifully summed up much of what I have learned through the world of construction. I mentioned that in my experience these skills carried over into family relations, and that I thought he was really talking about the very fabric of life.

David said, "I think that one of the big facets for me, the mediation training, it's the things I need to work out with my kids, my wife, my friends; these are important life goals."

Anyone who knows David knows how much he embodies these qualities and capacities we have talked about.

....

Tammy Lenski is another mediator/trainer/writer whom I respect tremendously, and appreciate and enjoy a lot. I first found some of her writing and blogging a few years ago, and right away I felt that Tammy has a great combination of wisdom and really good common sense, coupled with a deep capacity to observe others and herself. I felt she is well-grounded on earth while working with a big heart. Tammy is inspirational while at the same time full of practical advice. And she is damn _funny_. I find that typically the best mediators are funny. Humor disarms. Tammy can also be uproariously irreverent—another great quality, in my opinion.

I have not yet met Tammy in person, but we have connected through blogs and emails and now via telephone. Ahh, the plugged-in world. I am also starting into her book, _Making Mediation Your Day Job_.

Tammy and I had a rich conversation for this book, and she gave a warning about two things in particular that can stymie any attempt at open, clear communication and the building of trust. The first is someone being afraid to speak their truth even when cordially invited to do so. They might think something is too insignificant to bring up, or they might be shy or hesitant about bringing attention to it. Little by little, instances such as this can add up and gather quite a bit of energy, and in time turn into some sort of breakdown, _even though the people had been invited to speak forthrightly about anything whatsoever that concerned them._ As a result, someone might eventually build up some resentment and anger in reaction, and think the problem is with the other person, when the origin is actually their own hesitancy to speak up.

I became aware of this reluctance in my experience, too, and learned to encourage people to break through it by saying something like, "Now it is going to be much better for both of us if you speak your mind with me and tell me any concerns, questions and special wishes you have, or even what you might think is a crazy thought. It is better to get it out, and _nothing_ is silly to ask or bring up if it carries some weight _for you._ The more honest we are together the better we are going to do, and I expect to do really well together." I keep trying to put people at ease while drawing their truth out of them.

The other great point Tammy brought up regarding roadblocks to effective communication and trust is how we can form an impression of someone to the extent that we no longer see and hear them as they are. How we hold them is quite colored by our impression, which can become entrenched and solidified. This impression might be far from the truth, but it has *become* the truth for us.

I told Tammy that I want people to *risk honesty,* and to make strong efforts toward clarity and transparency and building trust. The conversation went on from there.

Tammy's powerful words in response included several great points:

"This approach is very similar to what I have felt and what I do with my clients, and what I frequently write about. I think the challenge in any project, whether it's construction or anything else, is that people do a couple of fairly normal human things that get in the way of that approach, even when they're making an effort to keep it really in the front of their minds.

"The first is that they tend to dismiss the things that they think won't really be a problem. So they may have a little worry, but they don't raise it, because they don't want to make a big deal. They don't want to be seen as aggressive. They don't want to be seen as a problem maker, so they shrug it off.

"What happens is that some of these things can add up, as you know. They add up and so each of those little things that individually doesn't mean much, those all add up to something that means a lot, and in time it's gotten so big that it's much harder to step in. Or, if something has gotten big enough that the people are frustrated, they may not feel motivated to bring their best to the conversation anymore.

"I think the trick with any kind of approach that asks people to step up early and address the things that their instinct is telling them are potentially a problem or run [counter] to their instinct, is to really also teach them that speaking up early—in other words, asserting their interests—isn't mutually exclusive from being *nice.* I find that people often think that if they speak up, along with it they risk being perceived as not nice anymore.

"So, that's the one thing that I think is almost a precursor to helping them take a—you're calling it a heart-centered approach—is helping them to understand that assertiveness and niceness are not mutually exclusive."

For me Tammy, as with all the fine mediators I know, is great at saying in a direct but non-threatening way, "Hey, the emperor is naked."

Here is the second big warning that Tammy noted:

"The other thing I think that people sometimes do is get stuck in what is called in systems thinking a *reflexive loop.* What it means is, if I meet you, and the first time we've met I listen to something you say and I reach a conclusion about it, I will understand it in a certain way based on my own experiences, my own beliefs, my own values, and so forth.

"So I form a belief about you or conclusion about you as a result, and the next time I see you, if we have a conversation, I am already bringing my beliefs and conclusions about you from the last conversation to the current conversation, and those feed what I hear you say and what I don't hear you say. So, I unconsciously effectively screen out things that might counter what I believe about you, and I equally unconsciously *let in* things that *support* what I think I know about you. It becomes a reflexive loop; it keeps feeding on itself.

"So the more I think, for example, that you're passive-aggressive, well darn, don't you start looking more passive-aggressive at every moment! *So I think that's also a challenge to the heart-centered approach—querying out those reflexive loops—and those are very hard, because of patterns of behavior that most of us have.*"

I want to note here that earlier in her life Tammy had experience with a contractor who would agree to do something and then just let it go. Eventually she realized he had no intention of doing what she asked, but said he would to appease her and, in that moment at least, make the matter go away.

"They're very normal human behaviors . . . It's something that goes ever so well when we really love somebody, because it could be a reflexive loop that's really positive, that is causing us not to see some things that we should see.

"But those two get in the way of speaking up early and speaking up often, because they essentially cause people to say things like 'It won't matter what I say, he's passive-aggressive.' It is, we believe, his state of being, in other words. So we disconnect our interpretation of him from what we actually hear him say. We don't believe that anymore."

Then, when it might be truly appropriate and in everyone's best interests for a client to raise concerns, someone may say, to quote Tammy, "It doesn't matter. He will be nice, and he'll say, 'Sure, that makes a lot of sense.' Then he's going to go do what he was going to do all along." And this sort of cynicism, coming from someone who is supposedly trying to look out for the best interests of his client, can have a devastating effect.

". . . But here is the challenge—that not all instinct is good instinct . . . Because if your instinct is informed by a bad reflexive loop, then it's lousy instinct potentially, because it's instinct that is incidentally connected with your harsh judgment.

"But I like the idea of risking honesty, not holding your cards as close to your vest . . . because when you are bargaining in a relationship that is ongoing, that can actually harm the relationship. We do it in the name if preserving the relationship, but it actually is going to have an unintended effect, which is to harm it."

At this point in our conversation I pointed out that, "One of the things I warn people about is assumptions, and I say the more you can let go of assumptions and really try to hear and see who you are truly dealing with, the better off everybody is going to be."

Tammy agreed, and added, "I think that people cognitively understand that assumptions don't often serve us very well. Even though we *have* to make assumptions day to day in our lives in order to navigate the world. We have to assume that the car pulling up at the stop sign on the cross street is probably going to stop, or life would be too scary to go out the front door every day. But at the same time, we have to hold a little caution in the back of our minds that *maybe* they won't stop."

This was another fine point from Tammy that some assumptions serve a helpful role in life, and even then we still need to hold open a little caution.

Furthermore, *I want to hold these two assumptions:*

1. *The world is a good place that is worth doing our best with, and*
2. *Being good to one another in our daily dealings helps the world person by person—and frequently brings surprising rewards of different sorts, not the least of which is a sense of expansion of our personal universe as we treat others with kindness, goodness and helpfulness.*

At the same time, I cannot assume that everyone will be on the same page and think like this!

We are talking balance and practicality here, and I do believe firmly that all our efforts to treat others as we would like to be treated build up like compounding interest, and help create true change. Gandhi said, "Be the change you want to see in the world." Heart by heart, little action by big action, it all adds up.

Tammy and I then talked about my recommendation that contractors and property owners get as much detail about a project as possible down on paper. This clarity is so important, I have found. A focused contract and a written record in the form of good specs serve all parties involved, and eliminate a lot of the possibility for disagreement and varying memories. Many of my customers have ultimately said how much they appreciate such detailed conversation, decision-making and recording.

Tammy responded, "That makes tons of sense. We think of memory as this accurate thing, as this concrete thing that 'what I remember is what *really* happened.' You and I both know that five people can be in a room together having a conversation and remember some very different things about that conversation . . .

"And it's because we hear things that have meaning to us, and we don't remember the things that don't have meaning to us. Furthermore, the more we tell ourselves a story about the conversation, the more that memory becomes ingrained in the hard wiring of our brain, and so we think it is just fact. We made it into "fact" because we kept telling ourselves it really is, over and over.

"That's what we do when we tell ourselves that about a conflict, and that's why when people come in and say, "Well, I want him to understand the truth about this," you and I know that everyone's truth is a little different in a conflict, and there is no one truth, and that subjective truth isn't very helpful.

"I loved something [else that] you said, and I wanted to come back to it . . . about having things in writing! I still to this day, when we have contractors coming to do something substantial, require specs—because I don't want to get into the conversation later about, 'Oh, gee! We understood you were going to use this kind of shingle, but you used that kind of shingle!'

"And I have found that really reputable contractors don't hesitate to do that. In fact, they are happy because it serves them well in the long run, too. And the ones who hesitate make me worry a little bit, that there is a reason they don't want to put that information down."

I heartily concurred, "Right—I totally agree with you, and I learned it the hard way in one particular project, and it had to do with my assumption of the husband's meaning about what the phrase 'storage unit' meant. So I learned, and now I say something like, 'a four foot wide cabinet made out of ¾ inch birch plywood, with adjustable shelving on the inside and raised panel doors, and three coats of high quality white paint.' I get very specific!"

The next comment from Tammy was priceless for me:

"I think we know that with our loved ones we can walk away from a conversation about choosing something, and we think we have made a decision together, and in fact each of us have a separate, slightly different version of what the decision was. We know that happens in our everyday lives with people we love and trust.

"So why wouldn't we think that could happen with people we barely know! It doesn't make any sense. So it's great that you are encouraging people to be detailed at the front end to help ward off problems later."

When I asked Tammy for some final wisdom, she said, "I guess the tip that comes to mind *is when you are most pissed at somebody, look for the equal human being in front of you! Not the person who is lesser in some way because they act in a certain way, but the person who is as smart as you and as graceful as you, and work with them from that place.*

"Look for what enables them to want to stop yelling, or what it is that's causing them to yell. It is not a psychological diagnosis of them, but a day to day, fair, usual reason that someone might yell—that might also compel you to do on occasion. Work with them from that place."

Compassion and humility.

This whole conversation with Tammy was both delightful and most insightful for me. She had added so many interesting and helpful pieces of insight from her experience and wisdom.

Tammy contributed one more very valuable slice of life that I want to share.

Earlier I told the story about having the late night conversation with my wealthy customer, in which I had finally just had enough of his attitude for the evening. I let him know that he had crossed over my line of personal dignity and respect, that he had gone too far. He knew it from my tone, too.

In a recent blog, Tammy wrote about a mediation session she had in which an otherwise charming, mature brother and sister were at odds over settling their father's estate, and they were both pretty foul-mouthed in front of Tammy and their attorneys. Lots of "f-bombs" were thrown about. It is clear that Tammy genuinely liked both people, and for a moment she was at a loss when the brother suddenly said he'd had enough, when they were in fact very close to a settlement—only five thousand dollars apart. He got up to leave, lawyers in tow.

At first, Tammy was really taken aback. In a moment, though, she knew what to do. She met the brother on the turf of his own language, asking him if he was "out of his f-ing mind" when they were so close to settling. Expecting the man to turn on her, she waited as he slowly turned toward her with a big smile, and agreed that he was in fact out of his mind. He then proposed to his sister that they split the difference and have dinner. She agreed. Done! Hours of wrangling solved with one charming and brutally honest question. That one cracked me up.

The point I want to make here is that I am not recommending "terminal niceness," and certainly neither is Tammy. I am not recommending wanton viciousness either. But sometimes a very blunt response is called for and can be just what it takes to shake someone else out of their stuck position.

The night I told my wealthy customer that he had crossed over my line I did not yell at him, but I had enough genuine emotion and firmness in my voice to get the point across. Tammy in her story said just the right thing, firmly and without malice.

....

For professionals in construction and related services:

My good friend, architect Richard Griffin from Salem, Massachusetts, takes his professional relationships to heart. I have known Richard for years. He is a genuinely caring and thoughtful person who takes his responsibility seriously. We have discussed the subject of *Conscious Cooperation* for years, and our mutual philosophies ultimately crystallized in one phone conversation I requested while writing this book.

We started by talking about clarity with clients. Richard said it is very important to have ". . . clarity in terms of what services we offer, what services the client expects, on what we are including and what our limits are." This area of communication, focusing on the need for clarity, has come up a lot in Richard's experience.

Talking of transparency Richard said, "That too is another important component: one party knowing what the other party is up to. Sometimes I suppose transparency can be overdone in terms of taking up more time to explain what one is doing . . ." when it is not really necessary.

On the other hand, "I think it is certainly a problem when people think they are getting the wool pulled over their eyes."

This last statement harkens back to the discussion with Tammy Lenski about *our perceptions of things.* If we have concluded that someone is out to pull the wool over our eyes, even if they are trying to be totally open with us we may remain stuck in our wrong assumption. This is where Tammy said it is important to try to clear the "reflexive loops."

Sometimes I ask brand new clients "Are you comfortable with me? Do you believe me? Do you have any reservations?" My experience, again, is that blunt but non-threatening questions tend to be disarming and build trust.

For a few years I worked with a very good, highly educated painter who said that for some time after we started working together he felt I kept waiting for him to exhibit weird behavior that he thought I must expect from all painters. I kept telling him this was not true, but for him it was. I told him that I had quickly decided he was good and honest, but it took quite a while for him to believe me. In fact, he had been a school teacher and was a part time music teacher and performer, an interesting guy. I think he was primed, though, so to speak, for people to expect to see stereotypical behavior often associated with painters. Stuck in a persistent reflexive loop.

Richard also hit upon a subject that comes up all the time. As Richard put it, "An issue I ran into most often relates to somebody not understanding why their bill is what it is. I ran into this more with managing construction contracts and not just doing the architectural work. I usually charge on an hourly basis, and sometimes things will take more hours than I originally projected, and I try to be frank about the problems that have been engendered. Very often people have not been willing to pay for extra hours."

I generally do construction work by the hour on a cost-plus basis, too—raw costs plus a contractor's percentage fee—and I have learned to stress to people up front that I have given them my best, honest estimate. However, estimating is not an exact science, unless you are repetitively doing the same thing or doing something that can easily be broken down into a unit price. I tell them the final cost could vary from my estimate, and that I will do my best to stay very close to the estimate. I ask if they understand this, and I write this into my proposals, too.

Here I think Richard summed it up quite well:

"I think overall if a client understands how a service provider goes about making their money, it doesn't have to be a mystery.

If they understand that a certain amount of profit is a legitimate thing, and this is the way business is done, [things are clearer from the outset].

"We tell clients what the overhead is, what the profit is, and this is exactly how we make our money. I frankly prefer to work that way . . ."

Previously in this book, I told about my amazement when some twenty-five years ago a new customer told me that working by fixed bid is normally unfair to either the homeowner or the contractor, and he suggested a cost-plus contract on his custom home. He was the only customer I knew who had ever concluded this—and suggested it to me!—on his own. Of course, it takes mutual trust from both sides to work in such a fashion, but it certainly can work, helped greatly by that matter of trust we've addressed in depth already.

Another important statement from Richard that goes back to the very foundation of a working relationship was, *"It's problematic when one doesn't get the sense that there is mutual good intention."* How true! I do not want to participate in such a relationship. The development of mutual good intention and the recognition of it from all "sides" is so important.

Richard added, "I think there is a pretty strong feeling of well-being when something is really going to work out well. And then, uneasiness can be present for a variety of reasons . . . And it surely should be investigated until one gets to the bottom of that feeling."

Richard also brought up the important point of clients feeling that they are getting their money's worth. This goes both ways, too; builders, architects and other service providers need to feel they are getting paid fairly, too, as I also addressed earlier herein.

Richard and I have a mutual friend who is a retired organizational consultant. He once told me that every morning as he was getting ready for the day he would look in the mirror and say to himself, "Now you have to be worth your fee today." What a nice way to approach work, giving the best of yourself. The wonderful consultant, trainer and author Harrison Owen once said, in effect, "No offense to anyone, but I think that 90 per cent of consultants' fees go toward making a living for the consultants and don't really buy much [for their clients]."

The reason he said this, he said, was that in his experience most consultants came into their settings as if they had the needed answers and formulae and could just impart wise information. In stark contrast, Harrison's approach is to deeply involve the participants. He tells them THEY are the experts, and he is a facilitator. He says this is the only way change can stick, and work, because the power and responsibility really lie with the people involved.

I think there is great wisdom in this approach. It mirrors how I like customers to be involved in planning and working out their projects with me, and to stay involved along the way. This strengthens and broadens our relationship and makes it all more real for them.

....

Carol Dilks is a long-time real estate agent from Reston, VA. Carol and I have had many talks about construction, real estate and the human interactions so important to both. She and I had a "Q&A" session for this book, too.

Carol is a passionate person. She is HERE. The opposite of blasé, she cares a lot about life and others. In her past, she did on-the-streets social work, and started a non-profit organization to help children who have lost parents.

Carol brings this kind of heart into her real estate work as well. She said she is here to serve the best interests of her clients, not to just try to ring up sales for herself. She told me how she recently walked away from a lucrative deal she knew she could have made, but she would not have felt in integrity doing it.

"I sleep well at night when I feel I have made really good communications . . . I usually have a pretty good read of the clients. There is nothing worse than miscommunication that down the road causes a problem."

In relation to connecting with her clients Carol said, "It's important to me why they're buying this house. The more I understand about their motivation, the easier it is for me to find them a house and help the process along by being a little more of a support system."

"It is not my job to agree with them. It is my job to support them." Carol said this in reference specifically to her dealings with many young couples, whom she has often felt were really stretching themselves to buy an impressive home. She said that she might bring it up one time, to make sure they truly want a particular home and feel good about the purchase and the financial commitment. Then it is up to them. As she said, she is there to serve. Part of that service is clarifying important questions and perhaps offering the wisdom of her years of experience, but then the final analysis and decision are theirs.

She said, "I try to help them not buy beyond their means," but she makes it clear that the choice is theirs. She resists the temptation to take on any more of a parental role than that.

I told Carol it sounded to me like she operates very much according to her moral code.

"Is there any commitment more important than that? If there is, I don't know what it is. My values are my moral compass."

I asked her if she has any advice for contractors. She said simply, "Follow their moral compass."

We also spoke of *assumptions,* and Carol spoke simply and eloquently of how assumptions can absolutely shut down communication and plug one's ears to anything other than what they "know." Here again is what Tammy Lenski referred to as the "reflexive loops."

Carol told me the story of how she was once helping a high-powered executive and his wife look for a house. Carol had a strong negative reaction to the wife. She could not stand her. As the day progressed, Carol was shocked to discover that her upset and dislike actually came from a great *similarity* between herself and the woman in one particular respect. The sudden realization was stimulated when she overheard a certain statement the woman made to her husband.

Because of this one statement a huge internal mirror was held up to Carol for herself and her own life, and the woman never knew it. To Carol's great credit she faced what she saw. The next day she and the wife went house-hunting without the husband. They did great together and found a house.

I love this story and think it is a wonderful illustration of the power of strong reactions to others, the potential sources of those reactions, and the major shifting that can come from facing the reactions squarely, taking responsibility for ourselves.

Another fine real estate agent in the same area, Shirley Mattam-Male, works with great commitment to her clients. I saw this first-hand when she listed and sold my friend Daniel's home.

Her priority was Daniel, and what she felt from her experience it would take to sell his home. I had previously done quite a bit of renovation to the house and did more renovation and service based on Shirley's recommendations. Shirley was excellent if exacting, and a buyer showed up the first day the house was listed. Shirley's commitment to excellence and success for her client stood out like bright lights. She did a great job, and her confidence was founded in her successful and observant experience.

....

Another interview for this book was with one of the planners of a large, medical complex. When I met this woman, Susan, she was involved in the planning stages of this major undertaking.

The same type of principles of human interaction and working together with positive intention and clarity apply, even though the field of large commercial construction is a world outside of anything previously discussed in this book. It is proof, to me, that it always does come down to *people*, and the more we can help people work as a team pulling together in the same direction and—one hopes—enjoying themselves while doing it, the better everything goes.

The following discussion also brings up the need to tailor the *process* of bringing people together in a collaborative way so it fits appropriately with particular projects. Again, the basic principles and intentions are always the same.

There are many more people involved, often in more complex ways, than with residential construction and remodeling work. A major hospital structure is a complex beast, especially with all the technical components that have to be designed in and installed correctly, in concert with many other components. A great deal of careful planning and design are called for.

Susan fully agreed that all key people involved in the design and construction process should be brought together ahead of time to establish open working relationships, and channels of communication for dealing with the challenges, changes and disputes that may arise.

She also pointed out that it is crucial for the "end users" to be involved at the time of design. How will they actually work in and maintain the complicated new buildings? In her words:

"By bringing everyone together in the same room at the same time—and that can include your end-users of your building, for instance, and your facilities management people—you can help the whole construction process. Usually that should result in a reduction in abrupt change orders and all of the associated costs for them, because you've got the various stakeholders involved from the start instead of being brought on board once the decisions have already been made. That should optimize the process and avoid having, for instance, building components and interior finishes and decisions around equipment needs that aren't very feasible from the lifecycle perspective."

My good friend Daniel Stone is an organizational consultant who has worked a good deal with the healthcare industry in various capacities. Among other things, he helps hospital personnel clarify their management needs and daily function needs so they can set up the most effective operations programs possible, to make optimal use of the hospital facilities. In addition, much of his of his work is also geared toward helping these personnel work effectively and collaboratively as a team. This team building is crucial, he says.

I told Daniel what Susan said about involving daily staff and even maintenance and housekeeping people in the design process of the hospital, and Daniel quickly agreed. He has done quite a bit of training and held numerous retreats for hospital personnel, and he said that the most important people for internal, functional design from his perspective are the nurses.

I emphasize this point because I find the subject of planning the design of a hospital and then building it both fascinating and so complex. *And, once again, at the core of planning, design, and construction, the best approach proves to be a collaborative approach that allows for input from all important constituents. It makes for a better facility and a more successful project. Once again,* it is still all about people working together with a clear, shared intent and a united spirit.

Susan also told me, "You asked about how making sure that the key players have plenty of time together ahead of time, to state their needs, wishes and intentions, would benefit the flow of the project. From where I sit, I think it is very important to team building in developing trust and establishing a healthy means of timely communication. It also helps to build bridges between the various professions.

"Sometimes, in part because of the way project management has been conducted in the past, they somewhat unfortunately pit some of those professionals against each other, because decisions are made in one realm of one profession that soon may impact the work, if not the timeline and the costs, of associated professionals. And so it can kind of get peoples' backs up if a decision has been made that impacts them, and they could have or feel they *should have* had some say on it but didn't.

"So I think it is probably very important to [establish early communication with key players,] to establish a team spirit and develop trust. That's really the basis for successful working relationships, as you know."

This early development of trust, a key component of successful working relationships, is increasingly crucial the bigger and more complex the project. Conversely, the greater the ramifications in problems and cost overruns later of NOT having done this.

Susan pointed out another important ingredient, saying, "Then we have how important regularly scheduled meetings are to continue and build on this intentional awareness of one another and to mutual support. Again, I think that is crucial, because it allows for better coordination between the team members and between various phases of the project. It would also go a long way towards ensuring that the proposed design solutions, and the materials that are chosen for a particular project, are feasible from the construction contractor's point of view, because I'm sure that person often will have some valuable input as well.

"For example, a certain material may be specified, but the contractor knows that there is a backlog of six months to get it, or it is no longer available in the area, or we have to get it from another country, which means more time and additional duties, and so on. So this sort of open input and exchange between the contractor and the design team, established early, is also key.

"There is also the importance of providing regular opportunities through those ongoing meetings to clear the air when conflict arises. I think it is highly important to establish a common purpose, because large projects are extremely complex, and they require very timely and very precise communication all the time, and there will be conflicts. It is highly important to be able to present conflicts and whatever else may get in the way of the flow of the work process."

We talked about the power of recording planning conversations, and using memos and notes that are available to everyone.

Susan commented, "Again, I think that's very important to provide written records of discussions, whether it's minutes or memos or whatever, and to invest in clear communication strategies. I think it's really important in reducing the chances of causing misunderstandings and bad feelings.

"*Generally, it is probably true that paying close attention to process is not necessarily how it is traditionally done, and there's a lot of assumptions made as to how things happen, kind of implicitly without explicitly stating objectives, means of communication, sorting out conflict and all that.*

"You asked how important it may be to make efforts to build and focus team spirit in a common purpose at the beginning. Again, I think it is essential, and it's all the more important because a functional team is largely about chemistry, and fostering that with crucial ingredients like trust and good communication.

"*So coming to a project such as this one with this kind of approach I think will help to keep people focused where it really needs to be to get the job done, because we're all going to have to pull together in the same direction.*"

Susan said that the *Conscious Cooperation* approach of having intentional collaboration is a natural step from the interdisciplinary design process commonly known and used now. This is what Susan is accustomed to. She also understood that we were talking about going a step or two further toward working together even better, because we are talking about truly recognizing the needs, wishes and natures of everyone involved, AND working to focus the best in everyone toward a large common purpose.

Of course, not everyone can have things as they would like all the time, but even if you can't have it all, a minimum effort toward truly VALUING everyone involved is extremely powerful in and of itself. When people know they are valued and appreciated, a door opens through which they can contribute the best of themselves. Everyone has an important role to play.

We are talking intangibles here, but they are intangibles that cannot be ignored and yet historically are often given scant attention.

....

So whether you are remodeling a bathroom or building a hospital, it still comes down to people, and how to make the most of the project and not just suffer through it with a preordained attitude that "everyone knows" how hard and miserable it will be!

Our attitude and expectations are huge.

The line from the Bible, "As a man thinketh in his heart, so is he," is so simple and so profound. The line is from Proverbs, chapter 23, line 7. In 1902, James Allen wrote a wonderful short essay based on this line. His essay continues to inspire people. As I find is often the case, deep truth is short and in a way simple. It is stripped down to its essence, and does not need tons of explanation.

For me, whatever religion you may follow, or not, does not matter at all. I believe we all have the same human DNA in us, with a presence and inspiration bigger than any of us that runs through all of us.

Even though this theme has been touched on in various ways in this book, somehow my conversation with Susan more deeply awakened my awareness of how critical are our attitudes, our degree of openness to others, our intention to give the best of ourselves that we can, our commitment to positive outcomes, and all the other aspects of *Conscious Cooperation*. I think somehow picturing that massive hospital project registered with me more starkly than any other just how different the experience and success of it could be if there is a strong basis of intentional collaboration versus "business as usual."

....

At this point, I would like to add some input from two friends who have both worked in construction for years. Both are very intelligent and have high Emotional Intelligence at the same time. Both are good with people in their individual ways.

Stephen Whalen is a carpenter and contractor from Cape Cod. He has done a wide variety of work over the years. He is a very pleasant and low-key guy, and you know he observes what is going on. Good radar.

I asked Stephen how important he thinks relationships are in his world of construction. He said, "Building a relationship with different customers, especially if they are going to have different projects for you, I think is key." He looks to develop a healthy, ongoing interaction with people.

We also talked about the matter of chemistry. Stephen commented, "Sometimes, you meet with someone, and you don't get a good feeling. Other times you get a good feeling right off the bat, and you know you can work with them."

He added, "I try to listen for what they really want. If there are two people involved, I try to get both of them there with me." We talked about how with couples this is especially important. We both acknowledged that old core skill of listening more than you speak. Stephen also talked of gradually building trust together, and to me that is a direct extension of listening, since the feeling of trust often is based largely on feeling listened to and sincerely paid attention to.

....

Another friend in the field whom I respect a lot is Steve O'Neil from the South Shore of Boston. Steve is also highly conscientious and wants to take care of his customers in the best possible way, even if that means that Steve is not the one who gets the job. He is serious about this. He has talked himself out of work before, telling people things like they would probably do better with a larger contractor who could get their particular work done more quickly. This is a pretty unusual stance for a contractor.

Not surprisingly, he has won some fans because of his sincerity, level of care and conscientiousness. He has a long list of grateful customers.

In his own words:

"Even if I don't get a job I leave the people with a suggestion that they get in someone else to do the work if for some reason the job is not a good fit for me. I want to make them and their house the priority. In this day when work is tougher to come by, a lot of contractors are taking anything, even if they are not really qualified."

Not to point any fingers, but I do agree with Steve that this happens to some extent anyway, and it dovetails with my advice to be honest. Don't try to sell yourself into something you are not qualified for! And on the other hand, don't be shy about stating your capabilities with confidence.

Steve added:

"Clarity and honesty are the crux of the relationship . . . with open communication, trust quickly builds. The customer can get to know you and gain new information."

Steve has strong ethics and very high integrity. He spoke of one man he met with just to look at work. The man insisted on giving Steve money for his time and advice. Word later got back to Steve that the man was raving about him at work. Steve said he "savors" such responses, knowing that "people are relying on me for who I am, the work I do and the honesty."

Steve further said that going the extra mile becomes part of his reward in his work and in life in general. Nice values! More proof of the point that David Hoffman made that not all rewards have to do with money.

Another time Steve drove a significant distance in pouring rain to look at a problem for a potential client. The man forgot he was having a large dinner party at the time he had scheduled Steve to come. Steve showed up at the door soaking wet, and when the guests heard who he was they enthusiastically invited him to join them in the food and drink. He said the owner looked on like a proud father. Clearly, the owner had spoken highly of Steve.

Like Stephen Whalen, Steve also said he listens for what people really want, which he often finds is not what they originally stated. Through further questioning and discussion, he frequently uncovers more information about their vision and preferences that changes the scope of the work.

Steve attended the daylong workshop I co-facilitated with David Hoffman and two other fine people in 2008. Steve was so enthusiastic about the day's content that he went and acquired the same excellent basic mediation training that I had received. He has been taking steps to make mediation part of his working life. This added facet to our friendship is very rewarding for me.

Steve said: "In general I believe in conscious cooperation. It is needed more, and there is enough work for everyone. It is the miscommunication in our industry that really drives everyone crazy."

He added: "I think that *Conscious Cooperation* can be applied in the corporate world as well. Communication will be improved, things will be done more effectively, and humanity itself will be helped. It can be applied in many different fields. That is why I am excited about being involved in the world of mediation."

Steve is a very giving human being who is eager to make more of a difference in the world, and I'm quite certain he will do so.

....

Customers

Ambrose and Anna are a couple who define dignity. If ever there were two people with strong moral fiber and a high regard and respect for others, it is surely them.

I did some modest work for them in the mid 1990's. They were very pleased with the results, and we got along fine. My dog Simba also charmed their son, who used to walk him in their neighborhood after school. Simba was the best PR agent I ever had.

Last year Ambrose and Anna wanted to do more work on their house, before they retired. They eventually tracked me down through my former wife, who accidentally gave Ambrose one wrong digit for my phone number (and yes, it really was an accident).

When he and I did finally talk on the phone, I envisioned their plans as less extensive than they eventually turned out to be. I ultimately ended up working at their house for several months.

They lived in the house the entire time, except for one trip that they took together.

The work included gutting the primary bathroom, gutting the kitchen, taking over the lower level family room for a while, changing interior doors throughout the house, sanding and refinishing oak floors, making considerable changes to the laundry/ utility area, installing an upgraded septic system, and more.

At the beginning of the project, I told them they would probably be doing themselves a favor to find another place to stay during certain stages of the work. I was concerned when they said they would rather stay in their own house and get by. They had owned the house for many years, and were willing to cope with a bit of chaos in order to stay in their familiar surroundings.

Well, we all pulled it off with no fur flying, and a lot of credit goes to their patience and cooperation. At the same time, I and all the men working with me took pains to respect their home and their daily life there; we made it work together.

Ambrose and I had a great conversation about working together.

I asked Ambrose why they had tracked me down after all those years for the second project. He said:

"We were very pleased with the project the first time, and working together was successful for us . . . and it goes beyond being pleased with the work. You were honest, along with providing good workmanship. So, I'd say it was both your workmanship and the integrity factor."

Regarding the most recent project last year, Ambrose said:

"Your personality is one that appeals to me, and I think to Anna as well. We were very pleased with the quality of the work this time again, and liked working with you as a person. You knew what you were doing and were certainly trustworthy.

"You helped us make decisions on selection of products and materials, and that was important. You made sure we had appointments to look at displays and samples, and to talk about our interests. You helped us make selections, like trying to make the best decisions for the kitchen countertops, and I really like how that worked out and how it came about. You took the time to go with us to meet with the salesperson and not just leave us alone with him. This was very productive and very important in the contractor/client relationship in terms of getting what we really wanted. And every skilled person you brought into our house to do a job was someone we would *want* to have in our house."

I then asked Ambrose how important he thought the human relations side of construction is. He replied:

"For us to engage with someone to do work for us, we've got to feel there is a trust factor and a personal relationship and that we are just plain comfortable with them. To us this is very important. It is not just cut and dried, 'Here is a glib contractor who says, "Okay, I can do this and that,"' and you just check it off the list. You can come up with projected plans and schedules that look good on paper, but we know that everything from weather conditions to environmental conditions can get in the way of meeting goals and objectives and timelines.

"You have to have someone who you don't feel you are 'getting the business from," and that there really are sometimes uncontrollable things standing in the way of progress and presenting unforeseen challenges. And you have to have confidence that the person you've hired will still carry on the best they can in the face of setbacks to produce the work expected. And, we were living in the house the whole time, too, so that factor made it even more important that we felt comfortable with our contractor. We had to feel comfortable with him being in our home so much while we were there."

At this point, I said that I definitely felt like "part of the family" for some time during their project.

Ambrose responded:

"That's pretty important. It's important to have a personal and comfortable relationship with an individual when you are working with so many decisions. Especially when your living space is being rearranged during the project, to become something you will be living with for some time to come."

There were other powerful aspects to our relationship together. One was that we had people in common in our lives, who have been very important to Ambrose and Anna. Another is that Ambrose and Anna happen to come from different racial backgrounds than I do. They discovered that I had done social work and community project coordination in black ghettos in South Carolina. Later we all realized that my former business partner and I had years before built the home of a prominent woman in Who's Who of Black Women in America, who happened to be a dear friend of theirs. And so on; the more they found out along these lines the more they felt affirmed that they were working with the right contractor.

Ambrose spoke directly to this:

"It's definitely important that you feel comfortable with the people working with you when you are from different cultures and backgrounds. It was interesting how you and I could end up in a relationship that had to do with our home, and we both care about the future of the world that we live in, and we can't pretend that the issues of different cultures interacting are not a factor in the United States."

For Ambrose and Anna, that comfort factor carried great weight. We do not always speak about these things in our day to day business relationships, and I am very glad that Ambrose, Anna and I could and did speak about them and bring them into our collective consciousness.

Feeling comfortable working together was certainly mutual in this case, and I knew that I was welcomed into their life.

CHAPTER FIFTEEN

In Conclusion (Bringing Out the Best in Yourself)

A few years ago I came up with the name "Conscious Cooperation."
My original website designer, Dawud Miracle (yes, that really is his
name), also happens to be a really good consultant. He loved
the name and encouraged me to use it. Simple, to the point, no
frills.

Since then, Ken Cloke said he thinks the more apt term is
"collaboration" rather than "cooperation." I think Ken is right
in the strictest sense; this is about working together with an
intentional, generous attitude. But I think that for most people the
word "cooperation" perhaps evokes an image more quickly than
"collaboration."

The other thing Ken said is that underlying what I am doing
and promoting is, in addition, a strong spirit of commitment and
openness. Again, that is both true and intentional.

While ultimately I decided to leave the name as-is, my thanks go
to Ken for another of his usual fine insights. There is more behind
the name than might be apparent. Very true. As I said before,
I have a deeper intention of contributing what I can to a world
that is a kinder, more open-spirited place, responsive to the needs,

wishes and realities of others. Here Ken's word "collaborative" applies well.

I believe deeply in the power to choose to be helpful, responsive and responsible. I believe in the power to choose to give the best of yourself and to work as a team when called for. I believe in the enormous power of acknowledging appreciation, as Kathy Stone stressed. Our power to choose in these ways is truly infinite, and profound. We can choose to be cooperative and collaborative, as we do the best possible work. This conscious choice can have a life-altering effect, in our lives and the lives of others.

I have also seen many times where such intentional mutual choices to work together and simply BE together in a committed, collaborative way appeared to create a magical atmosphere. Such an environment takes on a life of its own that is bigger than the sum of the parts. I do not know how else to describe it. Some kind of contagious magic appears in which people cannot seem to do enough for each other. There is enjoyment and fun as the people involved share their great work and accomplish an enormous amount together.

I think Ken Cloke and David Hoffman refer to this surprising state in some of their stories. It may be spiritually logical and an extension of *"as you sow, so shall you reap,"* but the *experience* of it can be quite delicious and transformative. Once you have tasted that I think it is hard *not* to live that way.

There is much written along the lines of self-help that encourages us to meet ourselves at a deep, still place within. To decide to take honest responsibility for ourselves and our actions, and to concentrate on the positive as we pursue our deepest dreams.

A lot of wisdom also has been shared about our ability to focus our minds and hearts to make great changes in our lives, through repeated energized positive thought and commitment. Sometimes I am truly amazed at how tricky our minds can be, and how easy it is at times to simply shake yourself off like a dog after a bath and plant a new, positive thought or affirmation into the old noggin and proceed ahead.

It goes both ways, too. It is so easy to get a negative, harmful or "default mode" thought going, often before we have a handle on it. Then our minds can be off to the races in a darker way.

It may be a huge challenge sometimes to even summon the gumption to try and stay on the bright and open side, and not give in to old messages and stereotypes. Nevertheless, I am convinced that our efforts all go to a positive repository and do have an effect. No positive effort like that is ever wasted. And it does take ongoing, evolving self awareness.

Mac Anderson, the founder of the business *Simple Truths,* said that many people seem to think that he is always upbeat and cheerful. He said that, to the contrary, one reason he founded his very successful business that shares positive messages is that *he* needs the messages himself.

I recently read that somewhere in the United States a woman finished her dinner in a neighborhood restaurant and asked for the bill. The waitress said that the couple who had just left had anonymously paid her tab. Astounded, the woman then quietly told the waitress she would pay for the couple near her. This "pay it forward" went on for two hours, with everyone paying for someone else, healthy tips included!

I am sure nobody who was in the restaurant will forget that evening. Such simple, daily miracles are to me what will change the world more than governmental actions, laws, programs or edicts. When hearts get touched real, profound and lasting change can occur.

Kindness, thoughtfulness and compassion are contagious, just as negativity can be. The tongue-in-cheek phrase "committing random acts of kindness" comes to mind and captures it well.

There are plenty of old sayings that reflect this same philosophy in different ways. "A smile on the outside turns into a smile on the inside." "We have the power to change our thoughts." We can "pick ourselves up and dust ourselves off" after a tough challenge, and even choose to face the challenge itself as a positive experience that will probably leave us deeper and bigger as human beings.

For me, and many others, there is much wisdom here. At the same time, sometimes it may be quite tough to carry through in such positive fashion, and we may need help. However, we *do* have the power to choose our attitudes and how we treat others, which applies to construction as much as to anything.

It is not about just getting through or getting by, or coping. The potential to live more fully and keep discovering new levels of ourselves and new levels of positive interaction with others is endless.

The deeper we go on this inner journey, the more aware we become of effects of our behavior and actions on others, and the effects that others have on us. We can choose to focus on the positive and share our light with others.

The inspirational writer and speaker Og Mandino has contributed elegantly yet simply to this body of truth. Napoleon Hill and James Allen, the latter already mentioned, are well-known inspirational writers from days past whose wisdom and application of it are eternal in their usefulness in our lives. Benjamin Franklin has also been widely quoted for his guidelines for a successful and happy life. There are many others from relatively modern times, too.

The wisdom is always there. It is expressed and applied in unique ways in different times. By now, there is a very large library available.

I have not mentioned great spiritual figures known to us from all over the world who have inspired and guided so many, and continue to, since way before our time. I do believe that the Golden Rule and the Ten Commandments, to mention some of the most widely known spiritual guidelines, apply to all of us and contain within themselves an infinite field of guidance and fulfillment. The major religions of the world repeat the same truths in different forms, and one of the most important is to treat others well.

Along these lines, somewhere during my career as a builder and a carpenter, I got the inkling that I could do more and have a richer experience in my business if I decided to be a more committed helpful and attentive person. I knew this might also mean taking on more responsibility as I stepped in more fully; I had to stand behind and deliver on what I said and probed into. In fact, this greater responsibility became true at a level I could not have anticipated. I had to be prepared to respond to the answers to my questions to clients, subcontractors and anyone involved in a project.

With this greater level of responsibility I embraced came a greater and greater awareness of detail in my craft and in my attentiveness to my customers. I became increasingly fussier as a builder, which did not hurt my reputation, either! The more I stepped in, the more my credibility and confidence grew, and the more my skill level grew. I had to embrace the standards I wanted to work by and have others work by, even if at times my attention to little details seemed excessive to some people.

With all this evolution, a richer life emerged. My reputation grew. Relationships deepened. I was recognized and involved in ways I never dreamed. I became more involved in my community. I came to see myself through an expanding lens, too, which was another unexpected effect. I allowed in the comments others made to me about myself. I observed the changes in how people responded to me as I embraced my role more and took more risks with the signals and inklings I received regarding the people in my work life, as well as in my personal life. This opening became a seemingly endless circle which fed itself.

It was about construction and much more. It was about maturing as a human being and doing my best to live the values being uncovered. It was definitely learn-as-you-go, and it still is!

The principles and life observations presented in these pages are not necessarily new, but perhaps how I have packaged them and probably how they relate to my experience is unique for me. That is true for all of us. I do value the uniqueness of everyone and believe that everyone has some kind of gift to give. Some gifts are bigger and more far-reaching than others are. That is all right. *We all have gifts.*

Uncovering and sharing those gifts is personal and unique for everyone, too, but there are some shared guidelines and commonalities. It is impressed upon me more and more every day just how powerful the mind is. If the heart guides it, it is a fantastic tool. If it takes over too much, watch out!

It is so easy to pass judgment, to stick to old assumptions and conclusions and to choose not to take an extra step toward one another, and to get caught in the "reflexive loops" that Tammy Lenski spoke of. That is a pity, and it is so human.

We are not always going to get it "right," and the dance on this earth may not always be beautiful and graceful, but we have the power to choose repeatedly to be the best of ourselves that we can, and that is an infinite doorway. The potential is truly unlimited, and we can decide how far we want to go. Every day is a clean slate.

In a sense, I did not have a choice that I was drawn to keep exploring construction relationships and to recognize how important these relationships are in construction and everything else, for that matter. The natural urge was there, and I guess that how far I wanted to go with it and what I wanted to do with it was up to me. Even there, maybe I did not have a choice, because I would have always regretted not taking this road and following it to places unknown. Call it destiny. There was a yearning that simply would not be quiet.

To put the message in different words, I have certainly concluded that we can make a choice to be collaborative and to think in terms of mutual success in construction (and life). It is not "us against them" unless we choose that. In addition, if your personal radar sends all kinds of warning signals to you regarding working with someone, on either "side," please listen to that! Either check things out further or respectfully part ways. Why make life more complicated than it has to be? On the other hand, if your radar picks up wonderful things about others, enjoy it!

At the suggestion of a friend a few years ago, I wrote two, two sided Guides for Successful Construction Projects, one for property owners and one for contractors. They are available through my website. This book is a big fleshing out of those Guides. The key points are the same.

Here is a brief summary of my critical advice, with a few tweaks from the advice presented in the short Guides:

1. Do your homework and be prepared, on both "sides."
2. Use your intuition about working together. Are you comfortable together? Do you feel enough trust to sign an agreement together?

   Be as open as you can be, while still holding to your integrity and listening to your personal "radar."

3. Check each other out!
4. Make a commitment to be a team and to give the best of yourself, whatever your role. This includes showing appreciation.
5. Be clear, be detailed and put job details on paper! Write a clear, concise contract that covers what work to be done, what are the specs, who is responsible for what, what is the money changing hands, what is a payment schedule,

and reference everything to dated plans and specs. And be careful about the dates, especially with revisions! I learned that the hard way, too. Make sure everyone is signing in reference to materials with the same dates.

6. Be prepared to deal with work change orders and unexpected occurrences. Construction is often by nature kind of a moving target. Address from the start how you will deal with changes.

7. Agree to communicate regularly.

8. Speak up! Do not let questions and annoyances fester.

9. Do your best to hold judgments and assumptions at bay. *Listen* as openly as you can. Beware of thoughts and statements like, "Oh, all builders . . ." or "All homeowners . . ." No, they are NOT all . . . fill in the blank! Are you just like everyone else? There really is not an "everyone else"!

10. Use humor together. Be prepared to laugh and smile. It is disarming. Take yourself as lightly as you can, while also standing for who you are. Again, Tammy Lenski said it so well, that being nice and also assertive at the same time don't have to be mutually exclusive.

11. Agree to try to work out any disputes together, and if you need outside help, please turn first to a reputable mediator. You can turn to me, if you like.

12. Perhaps the most important tip: Enter the relationship with a commitment to mutual success. If you follow everything else I suggest and then apply this step, you can EXPECT success. It is also very powerful if you PICTURE mutual success. The vision that you hold has real power to it.

Remember, if you expect success and make your best contribution to it, as long as you feel good about each other, you SHOULD experience mutual success. If you expect difficulty or disaster, you sure increase your chances of experiencing it. <u>You attract what you give attention, pleasant and unpleasant.</u> This is a major core of the Law of Attraction.

13. <u>Go for the passion!</u> I am just adding this one. In her charmingly to-the-point book <u>Making Mediation Your Day Job</u>,* Tammy Lenski makes a wonderful statement. Actually, she makes many, but here is the one I want to share right now: "If you're a mediator who hesitates to convey your passion for what you do, you've taken the concepts of neutrality and professional distance way too much to heart."**

To me this applies to everything, including construction. If my truck needs work, I want to take it to a mechanic who loves trucks and cars and cares about MY truck, not a guy who is punching a clock and could not care less. If I need a doctor, I want one who <u>cares</u>, is compassionate, and holds himself or herself to high standards. I had some medical testing done last year, and the doctor had the most gracious bedside manner. One nurse told me they were all in love with him, and I was lucky to have him.

Recently I needed *two* emergency root canals at the same time. I was in an unfamiliar area and took a chance on a dentist. My gut said yes. When she finished the repair work, I thanked her for squeezing me into her schedule, even though I didn't exactly enjoy the root canals. That night around nine o'clock she called me just to see how I was doing! How many dentists do you know who do such a thing?

The same certainly goes for construction. I know that all my subcontractors CARE about what they are doing. They are committed to quality work and service. They want me to be happy.

_____
* *Making Mediation Your Day Job,* Tammy Lenski, i Universe, Inc., p. 16

They want my customers to be happy. We enjoy one another and working together. My customers have made many comments about this over the years, and they have enjoyed the process of our "team."

Even though the construction business has driven me nuts more than once, I CARE about my customers and the work I am doing for them. I want to work to a high standard even if no one is watching, and even if a particular detail will never be seen by anyone. I want to be happy in my work and in my interactions with others. I want to add something good and hopefully beautiful to their lives. I am not alone. There are plenty of craftsmen and women, and architects and interior designers, engineers, realtors and so on, who are made the same way.

Where I have been working lately in a new area I was referred a lady painter from Nicaragua, named Claudia. She is GOOD at what she does, she has high standards, a high level of collaborative spirit, and she is a pleasure to have around. She has a very healthy *passion in her spirit*, and I am fortunate to have found her. She exemplifies the qualities I have just described. I could tell her to paint a house for me and leave for the entire time, knowing that she would perform to a high standard and think well along the way.

The referral to Claudia came from a man named Martin, who also worked with me. with me. He knows a lot. I trust his integrity, and I am fussy. In some areas of construction he knows more than I do, and he knows there is always more to learn and a higher level of mastery to achieve. I challenge him to be the best he can be, and usually he likes it. Sometimes I may be a pain in the neck to him, but he knows I want to do things thoroughly.

Martin is very loyal, and he is a cooperative and pleasant human being. He and Claudia both care about the people they work for and with. This spirit is very important to me. I look for it. Martin cares about me. I am not just a source of work for him. It is nice, and it goes both ways.

You know when someone has high standards, and part of those high standards is knowing they can always learn. Confidence is indispensible, and so is humility. To some extent, we come into this world with a general makeup, and then what we do with it is, I think, an open book. Sometimes there are people raised in the bleakest of circumstances that achieve wonderful things and become human beings who are a gift to the rest of us.

I choose to believe what are, for me, truths such as these. They form a lot of the basis of Conscious Cooperation.

RESOURCES

Media:

My e-booklet Win-Win Construction Projects and my two "Guides for Successful Construction Projects" are available through my website:

www.consciouscooperation.com

Ken Cloke has numerous books and more available through www.kennethcloke.com Ken also offers workshops, and he is a fantastic mediator.

Tammy Lenski has different arms of her business, accessible through www.lenski.com She, too, is a trainer, speaker and excellent mediator

David Hoffman is the founder of the Boston Law Collaborative. He and his writing may be found through www.bostonlawcollaborative.com. David is a superb collaborative attorney and mediator. He also offers training sessions and teaches at Harvard Law School.

....

SERVICES:

I am available for mediation, consulting, training, speaking and coaching:

stuartbaker@consciouscooperation.com
888 518 0707

The fine people listed above may also be available. Local mediation services may be found in most areas through a little search. I always think it is best to get some strong recommendations for any services, including for mediators, as I recommended for contractors. There are frequently community mediation services, which might be able to suggest skilled private mediators, too.

These days so many professionals have a presence on the internet, so if you Google mediators in any area you should come up with a list. Again, I would still look for positive recommendations.

Another friend and very talented consultant is Brent Darnell from Atlanta, GA. Brent works largely in the world of larger commercial construction, offering emotional intelligence diagnostics and coaching and a total life approach to all-around health, happiness and fulfillment in collaboration with his wife. Brent also has publications available through this site: www.brentdarnell.com.